Ecohydrology-Based Landscape Restoration

This book provides an introduction to a fairly new approach to natural resources management practice entitled ecohydrology-based landscape restoration.

Ecohydrology-based landscape restoration integrates landscape restoration practices into ecohydrology science and principles in order to help address the limitations of current management practices in developing countries. Focusing on both the theory and practice of implementing new management practices, the book includes conceptual designs and practical demonstrations for a variety of sites, including hillsides, farmlands, gullies, riparian buffers and wetlands, while also drawing on field research conducted in Ethiopia. The book puts forward principles for improving current practices, which include the better integration of hydrological and ecological concerns, the greater involvement of local communities, the adoption of indigenous practices, the establishment of green and semi-gray infrastructure as an ecohydrological systemic solution, and the necessity of taking an adaptive approach to managing landscapes.

This book will be of great interest to students and scholars of water resource management, ecohydrology, and landscape restoration as well as professionals involved in the restoration of landscapes in developing countries.

Mulugeta Dadi Belete is Associate Professor of Hydrology and Water Resources at Hawassa University, Ethiopia.

Routledge Focus on Environment and Sustainability

Water Governance in Bolivia
Cochabamba since the Water War
Nasya Sara Razavi

Indigenous Identity, Human Rights, and the Environment in Myanmar
Local Engagement with Global Rights Discourses
Jonathan Liljeblad

Participatory Design and Social Transformation
Images and Narratives of Crisis and Change
John A. Bruce

Collaborating for Climate Equity
Researcher–Practitioner Partnerships in the Americas
Edited by Vivek Shandas and Dana Hellman

Food Deserts and Food Insecurity in the UK
Exploring Social Inequality
Dianna Smith and Claire Thompson

Ecohydrology-Based Landscape Restoration
Theory and Practice
Mulugeta Dadi Belete

For more information about this series, please visit: www.routledge.com/Routledge-Focus-on-Environment-and-Sustainability/book-series/RFES

Ecohydrology-Based Landscape Restoration

Theory and Practice

Mulugeta Dadi Belete

Routledge
Taylor & Francis Group

LONDON AND NEW YORK

from Routledge

First published 2023
by Routledge
4 Park Square, Milton Park, Abingdon, Oxon OX14 4RN

and by Routledge
605 Third Avenue, New York, NY 10158

Routledge is an imprint of the Taylor & Francis Group, an informa business

© 2023 Mulugeta Dadi Belete

British Library Cataloguing-in-Publication Data
A catalogue record for this book is available from the British Library

Library of Congress Cataloging-in-Publication Data
A catalog record has been requested for this book

ISBN: 978-1-032-31316-0 (hbk)
ISBN: 978-1-032-31318-4 (pbk)
ISBN: 978-1-003-30913-0 (ebk)

DOI: 10.4324/9781003309130

Typeset in Times New Roman
by codeMantra

To my loving and supportive wife *Simegn Asmare* who has committed herself to get this book published instead of demanding the care and time she so deserved by offering her never-failing sympathy and encouragement.

Contents

Contributors

Ayualem Ahmed is a Lecturer at the Water Resources and Irrigation Engineering department, Hawassa University, Ethiopia. She received her B.Sc in 2015 in Water Resources and Irrigation Engineering and her M.Sc in Water Resource Engineering and Management in 2020 from Hawassa University's Institute of Technology.

Mulugeta Dadi Belete is Associate Professor of Hydrology and Water Resources at Hawassa University, Ethiopia. He received his B.Sc. in Agricultural Engineering in 1999, and his Master's degree in Soil and Water Engineering in 2004 from Haramaya (the then Alemaya) University in Ethiopia. He had then received his PhD in Hydrology at the University of Bonn, Germany, in 2013. Dr. Mulugeta is a licensed practicing professional in water resources engineering. In this regard, he is half-practitioner and half-academician for being a university professor, an international consultant, and a project implementer at the lowest tier of government administrative division.

Markos Mathewos Godebo is an Assistant Professor of soil resources at Hawassa University, Ethiopia. He received his B.Sc. in Agricultural Engineering and Mechanization in 2005 and a Master's degree in Soil Science in 2010 from Hawassa University in Ethiopia. He received his PhD in Soil Resources from the Hawassa University in Ethiopia in 2020.

Abebe Beyene Hailu is an Associate Professor of Environmental Health and Ecology at Jimma University, Ethiopia and a Guest Lecturer at Vrije Universitiet Brussel (VUB), Belgium. He received his B.Sc in Environmental Health from Jimma University and M.Sc in Environmental Science & Technology and PhD in Science from VUB in 2006 and 2011, respectively. He has been conducting research in Tropical Limnology and received an award of Laurate in Tropical

Limnology research from Belgian oversea Science for his outstanding scientific contribution.

Yohannes Zerihun Negussie is the Director for African Regional Center for Ecohydrology (ARCE) under the Auspices of UNESCO, Ecohydrology Directorate in the Ministry of Water and Energy of the Democratic Republic of Ethiopia. He received his B.Sc. in Agricultural Engineering in 1987 from Haramaya (the then Alemaya University) in Ethiopia and his Master's degree in Ecohydrology from University of Lodz, Poland in 2012; and from Christian-Albrechts-University, German in 2013 under the Erasmus Mundus Program.

Bekele Beriso Sorsa is an Assistant Lecturer of Natural Resource Management at Dilla University, Ethiopia. He received his B.Sc. in Natural Resource Management in 2017 from Wachemo University, Ethiopia, and Master's degree in Soil and Water Conservation Engineering in 2021 from Hawassa University, Ethiopia.

Maciej Zalewski is the Director of the International Institute of Polish Academy of Sciences - European Regional Centre for Ecohydrology under the auspices of UNESCO; Director of the Department of Applied Ecology University of Lodz; and Director of the Center for Ecohydrological Studies University of Lodz. He is also a Chief Editor of "Ecohydrology & Hydrobiology" journal and a Member of Editorial Board for "Fisheries Management and Ecology", UK "Brazilian Journal of Biology", Brazil, & "Journal of Environmental Accounting and Management", USA.

Preface

The emergence of a rational systematic management of natural resources can be traced back to the late 19th century. Through dynamic progresses, we have now reached the decade of ecosystem restoration that recognizes it as a precursor for the achievement of the 17 Sustainable Development Goals. It is apparent that restoration of ecosystems has never been more urgent than these days, and we are already a little late to step up our efforts. While discharging these high-level global commitments, the ecosystem restoration initiatives should not be run as 'business-as-usual' anymore. They need to be supported by trans-disciplinary sustainability sciences such as ecohydrology. By relying on the principles of ecohydrology, we are using ecosystem properties as management tools toward sustainability of natural resources. This book is an output of such shift in paradigm to come up with a fairly new approach called 'Ecohydrology-Based Landscape Restoration' coined 'EcoLaR'. The book sews together the concepts of landscape restoration and ecohydrology for effective restoration of degraded ecosystem. The introduction of Ecohydrologic Systemic Solutions (EHSS) in the form of green- (semi-) gray infrastructure in the ecosystem restoration initiatives will assist the recovery of an ecosystem that has been degraded, damaged, or destroyed. By doing so, the restoration industry will be capacitated to be resilient to climate change and significantly contribute toward the betterment of community livelihood.

Acknowledgments

This book would not have been possible without the support of Hawassa University and Ministry of Water and Energy-Rift Valley Lakes Basin Development Office (RVLBO). Their collaborative effort in the modality of university-industry linkage was an instrumental to realize this book. The research part of the book is funded by the Office of Vice President for Research and Technology Transfer of Hawassa University under the thematic research project while the development and final write-up was sponsored by RVLBO within the objectives of devising an innovative tool for implementation of its strategic basin plan (2021–2035) and national guideline. The Center for Ethiopian Rift Valley Studies (CERVaS) deserves special acknowledgment for hosting the project. The Natural Resources Stewardship Programme (NatuReS) of GIZ and Norwegian Agency for Development Cooperation (NoRAD) also deserve due appreciation for allowing the university to experimentally demonstrate this fairly new approach of ecohydrology-based landscape restoration (EcoLaR) in their project areas through their private-public-civil society partnerships modality. My family members, especially my wife Sr. Simegn Asmare, passed through several challenges to make this book happen. The professional review of the manuscript by Prof. Tesfaye Abebe, Dr. Mihret Dananto, Dr. Abiot Legesse, Dr. Tekalgn Ayele, team members of the research council of CERVaS, and technical team members of RVLBO has given it the present shape, so I am very grateful to all of them. The editorial effort of Dr. Emebet Bekele also deserves due acknowledgment for enhancing the message-delivery power of the book.

1 Philosophical foundations for Ecohydrology-Based Landscape Restoration (EcoLaR) approach and limitations of the conventional practices

Mulugeta Dadi Belete

1.1 Introduction

Landscapes all over the world are diverse and play a crucial role in our lives (Machar, 2020). Landscape degradation, which is closely linked with water scarcity (UNCCD and FAO, 2020), has been well recognized as a major threat to human wellbeing and environment (UNCCD, 1994). Along history, natural resources in the landscape have been explored to satisfy human's hunger for land and thirst for water. Meanwhile, population has kept on increasing, and the developing economies have been getting involved in unsustainable production systems and consumption patterns that exert extreme pressure on both biotic (living) and abiotic (non-living) resources (Ariti et al., 2015).

Landscape degradation refers to the reduction or loss of the biological or economic productivity and complexity of land (UNCCD, 2016); it is thought to be in the order of 10–17% of global gross domestic product (GDP) (ELD, 2015). Landscape degradation directly affects 1.5 billion people globally, of which most live in extreme poverty (Agostini and Purdie, 2017). It has more severity in Africa where 46% of the continent has been threatened (WMO, 2005), and is still expected to worsen (IUCN, 2017). Such continuously pressing environmental problems underscore the need for a systemic understanding of environmental and social problems, and the creation of adaptive management and policy solutions (West et al., 2014).

The concept of landscape management, which dates back to 2000 BC, has been evolving and continuously improving over time (Chen, 2007; Khan et al., 2021; Zheng, 2004). However, up until the end of the 20th century, landscape management was dominated by a mechanistic approach with focus on elimination of environmental threats and little

DOI: 10.4324/9781003309130-1

regard for the impact of those practices on the ecosystem (Zalewski, 2021). As a result, ecosystem services faced serious threats and rapidly diminished in many landscapes around the world (Falkenmark, 2003) despite their beneficial value for families, communities, and economies (Boyd and Banzhaf, 2007) and maintenance of the conditions of life on earth (Deal et al., 2012).

The positive impacts of some of the prevailing landscape restoration practices such as the conventional soil and water conservation measures have been reported in improving soil physico-chemical properties (Mekuria et al., 2006) in: reducing soil loss (Taye et al., 2013); reducing sediment yield (Haregeweyn et al., 2006); conserving soil moisture (Vancampenhout et al., 2006); improving crop growth and yield (Teklu et al., 2018); regenerating vegetation and soil build-up (Descheemaeke et al., 2006); providing fodder yield (Kebede, 2015) and farmers' income (Amede, 2003); and so forth. However, in spite of millions of dollars invested in watershed management and landscape restoration in developing countries, the land degradation problems are still increasing (Osman and Sauerborn, 2001) and posing sustainability questions on the conventional approach.

Landscape restoration deals with large-scale processes in an integrated and multidisciplinary manner, combining natural resources management with environmental and livelihood considerations (FAO, 2012). It is primarily rooted in conservation and the science of landscape ecology (Sayer, 2009) which display a clear overlap with at least five of the key objectives of the Sustainable Development Goals (SDGs) (end hunger; secure water; promote strong, inclusive, and sustainable economic growth; tackle climate change; and protect and promote terrestrial resources) (Reed et al., 2015). However, in this rapidly transforming Anthropocene Era, it is apparent that landscape management requires the ability to move beyond typical reductionist approaches toward more holistic methods (Bunch et al., 2014).

The restoration of 'ecological' and 'landscape function' has become an important theme of recent scientific and policy work. It is not simply planting trees (Baig et al., 2017), but generally refers to the process of assisting the recovery of an ecosystem that has been degraded, damaged, or destroyed (SER & PWG, 2004), and it plays a central role in the provision of ecosystem services and the realization of the UN's SDGs (Yirdaw et al., 2017). It is more than land restoration (Sola et al., 2020) and usually targets the reparation of ecosystem processes, productivity, and services without necessarily achieving a return to 'predisturbance' conditions (CBD, 2012; Mansourian, 2005). In this sense, restoration is conceived as a triple win solution to regain ecological integrity, enhance human wellbeing, and resilience to climate change (Pfund and Stadtmüller, 2005); to be quintessential for the conservation of the threatened and unique biodiversity (Yirdaw et al., 2017); and to become a priority for combating desertification and land degradation and for limiting the impacts of anthropogenic climate change (Aronson and Alexander, 2013).

At this level of understanding, it is apparent that if the 'business-as-usual', that is, the sectoral and dominating mechanistic-deterministic, approach is continued, the existing socioeconomical gaps will deepen and realization of the SDGs will be undermined (Jarosiewicz et al., 2021). It is also expected that only a few decades are left before the declining functioning of the biosphere clashes with unattainable expectations of global carrying capacity, with consequences of local, regional, and global conflicts (Zalewski, 2014).

In this line of progress, different approaches were formulated such as: Forest and Landscape Restoration (FLR) (IUCN and WRI, 2014); Ecosystem Approach (CBD, 2004); Ecosystem-Based Adaptation (CBD, 2009); Ecological Restoration (SER, 2008); Protected Areas (Dudley, 2008); and the most recent Ecosystem Restoration (FAO, IUCN CEM &SER, 2021). However, the following methodological gaps are observed in all of these:

- Less attention is given to the concept of regulation of the fundamental ecological process, first and foremost –cycling of water and nutrients, which is a basic role of ecohydrology (EH) (Zalewski, 2021).
- The technical dimensions of the conventional landscape restoration practices have been scantly innovated due to the general perception of 'maturity' and 'availability' of the techniques.
- Enhancement of sustainability potentials (Water, Biodiversity, Services from ecosystems for society, Resilience to climate and various anthropogenic impacts and Culture and Education – WBSRCE) (Zalewski, 2021) are not explicitly targeted despite the fact that it is impossible to achieve sustainable growth without addressing these six dimensions (Jarosiewicz et al., 2021).
- The conventional approaches tend to have less concern for the concept of Abiotic-Biotic Regulatory Continuum (ABRC) (Zalewski and Naiman, 1985), which underlies effective ecological restoration of a given landscape.
- Their intrinsic logical paths of change (theory of change) need to be re-visited in the eyes of EH which is considered as the science of sustainability.

Those limitations heighten the importance of re-thinking how degraded landscapes/ecosystems were restored in a sustainable way that inspired and became a foundation of the proposed, fairly new approach, known as 'ecohydrology-based landscape restoration' (coined as EcoLaR), which can serve as a complementary method to the existing approaches. Particularity of this proposed approach lies in its tendency to ecologically re-engineer the existing landscape restoration practices with a motive to fine-tune some of their prevalent limitations. The attempt of re-engineering was supported by shifting the paradigm into ecohydrologic strategy which employs the water-biota interactions for environmental management, as well as researching, practically demonstrating, and testing of the findings in the real world.

1.2   Ecohydrology as a pathway to move from 'mechanistic' to 'ecosystemic' approach for sustainable landscape restoration

About 70% of the Earth's surface is anthropogenically modified resulting in the distortion of ecological processes propelled by water and nutrient circulation and changing the environmental equilibrium of the Earth (Zalewski, 2014). In the recent stage of Anthropocene, harmonization of human needs with the biosphere potential is the primary challenge for a sustainable future of the global ecosystems and the society (Burdyuzha, 2006). It is apparent that it is not enough to conserve the nature, but much more intensive action is necessary toward reversing degradation of biocenosis. For this to happen, there is an urgent need to change the mechanistic paradigm into Evolutionary/Ecosystemic in order to achieve sustainable future (Zalewski, 2021).

One of the potential pathways toward sustainability, which has been gaining a growing interest, is the use of potential for water/environment/society problem solving through scientific exploration of water-biota interactions (Wood et al., 2007). Here, we need to reconcile the two contradicting approaches to water resources management (hydrotechnical and ecological) within the context of EH (Zalewski, 2014). EH, the study of interactions between ecological and hydrological processes (Porporato and Rodriguez-Iturbe, 2013; Rodriguez-Iturbe, 2000), has developed rapidly in the past two decades in response to watershed ecological degradation amid environmental changes worldwide (Asbjornsen et al., 2011). It represents an integrated understanding of biological and hydrological processes at a catchment scale in order to create a scientific basis for a socially acceptable, cost-effective, and systemic approach to the sustainable management of freshwater resources (IHP, 2013).

Jørgensen (2016) justified the role of EH as an environmental management tool for its powerful and cost-moderate practices, while Zalewski (2021) considered it as an interrogative sustainable science. Jackson et al. (2009) also advocated it as a very much applied science with a focus on problem solving (Nuttle, 2002) and having a firm theoretical foundation (D'Odorico et al., 2012).

Landscape conservation and restoration must be a key element of our economy, but it is not (Brasser and Ferwerda, 2015). If the business-as-usual strategy continues, the existing socioeconomical gaps will be deepened and realization of the SDGs will be undermined (Jarosiewicz et al., 2021). In order to successfully respond to the threats of deepening socioeconomic gaps and for life to continue on Earth, these distorted ecological processes need to be restored, ecosystem functions maintained, carrying capacity and flux transference mechanisms should be enhanced using the basic properties of resilience as a water management framework (Zalewski et al., 1997). The science of EH has developed rapidly in the past two decades in response to these threats (Asbjornsen et al., 2011), as an important scientific field to address human influences on water resources and ecosystems.

EH was first defined by Zalewski et al. (1997) and considered as a sub-discipline of ecological engineering (Jørgensen, 2016) on the one hand and a sub-discipline of hydrology (Zalewski, 2014) on the other. In addition, Wassen (2007) viewed it as a landscape ecological specialization, while Jørgensen (2016) considered it a sub-discipline of ecological engineering. However, it is apparent that the science is shared by the ecological and hydrological sciences (Nuttle, 2002) with the notion of bridging between hydrology, ecology, and environmental management. In other words, EH emerged by the blending of theories of two physical sciences: ecology and due to their scientific overlap, and it signifies the impact of hydrology on ecosystems and vice versa (Zalewski, 2002).

In practice, EH is more than just hydrology and ecology combined, and it is proposed as the best practice for environmental management to achieve sustainable development (Zalewski, 2002). The essence of ecohydrologic strategy for environmental management lies on the possibility of regulating hydrological and ecological (dual regulation) processes, especially in novel ecosystems, as an alternative to conservation and restoration measures, to increase carrying capacity of a landscape (Zalewski, 2015). It suggests an integrated understanding of biological and hydrological processes at a catchment scale in order to create a scientific basis for a socially acceptable, cost-effective, and systemic approach to the sustainable management of freshwater resources (IHP, 2013). The underpinning paradigm shifts at the base of EH comprise: (1) the belief in an unlimited nature's potential; (2) the conquest of nature through agricultural revolution; (3) the exploitation of nature through industrial revolution; (4) the protection of nature through defensive environmental tactics in the 1970s; and (5) the restoration and regulation of the ecology. In this new strategic management of the ecosystems, the concept of 'regulation' of ecohydrologic processes appears to be the key element (Zalewski, 2013) to develop a methodology on how to regulate hydrological cycle at various spatial scales and formulation of systemic solutions for enhancement of sustainability potential WBSRCE of catchments (Zalewski, 2021). Identifying the hierarchy of factors that regulate the dynamics of the environment is the basic question from which the formulation of EH theory arose (Zalewski, 2010).

Here, EH can offer a scientific basis for designing more holistic and integrative approaches at the interface of hydrology and ecosystem science. It usually utilizes these feedbacks as a management tool in order to enhance quality of ecosystem, which represents a desired endpoint of environmental management (Costanza and Mageau, 1999). EH provides not only scientific understanding of the hydrology-biota interplay but also a systemic framework on how to use ecosystem processes as a complementary tool to already applied hydrotechnical solutions. Generally, EH provides three new aspects to environmental sciences and their integration into problem-solving mechanisms (Zalewski, 2014):

1   The need to use ecosystem properties as a complementary management tool harmonized with hydrotechnical solutions
2   The necessity to enhance ecosystem's carrying capacity by using the interplay of hydrology and biota
3   A requirement of regulation of ecological and hydrological processes as reflected by nutrients'/pollutants' absorbing capacity versus human impacts based on a 'dual regulation' approach

Thus, this chapter aims at synthesizing and proposing the guiding principles of EcoLaR approach in response to the top ten limitations of the conventional watershed management practices published by Belete (2021).

1.3 The key philosophical pillars of the anticipated 'ecohydrology-based landscape restoration' (EcoLaR) approach

EcoLaR approach is a fairly new approach to fine-tune the existing practices of landscape restoration with a major focus on terrestrial part of a given landscape by extrapolating the fundamental theories of EH. The key philosophical foundations of the EcoLaR approach is grounded in the following four concepts, where the first is related to resource redistribution pattern (Ludwig et al., 2005), whereas the other three are related to the EH hypotheses (Zalewski, 2000):

1 Feedback interactions between ecological and hydrological events and processes in a landscape logically follow the Trigger-Transfer-Reserve-Pulse (TTRP) chain.
2 The regulation of hydrological parameters in a landscape can be applied to control biological processes.
3 The shaping of the biological structure of an ecosystem(s) in a landscape can be applied to regulate hydrological processes.
4 Both types of regulation integrated at a landscape scale and in a synergistic way can be applied to the sustainable development of freshwater resources, measured as the improvement of water quality and quantity (providing ecosystem services).

The subsequent sub-sections describe these four concepts in detail.

1.4 The Trigger-Transfer-Reserve-Pulse (TTRP) framework to explain ecohydrologic interactions in the landscape system

Given the complexity of ecohydrologic interactions, frameworks have proven to be useful tools for conceptualizing and synthesizing complex interactions between landscape patterns and processes (Ludwig et al., 2005). The foundation of EcoLaR approach is linked to the basic understanding of how natural landscapes function over space and time to retain and use vital resources (Ludwig et al., 1997). An important conceptual advance in describing and clarifying the linkages between surface runoff and vegetation patches is the TTRP logical framework (Figures 1.1 and 1.2) (Tongway and Hindley, 2004; Ludwig et al., 2005). The framework enables us describe the links and interactions between ecological and hydrological events and processes

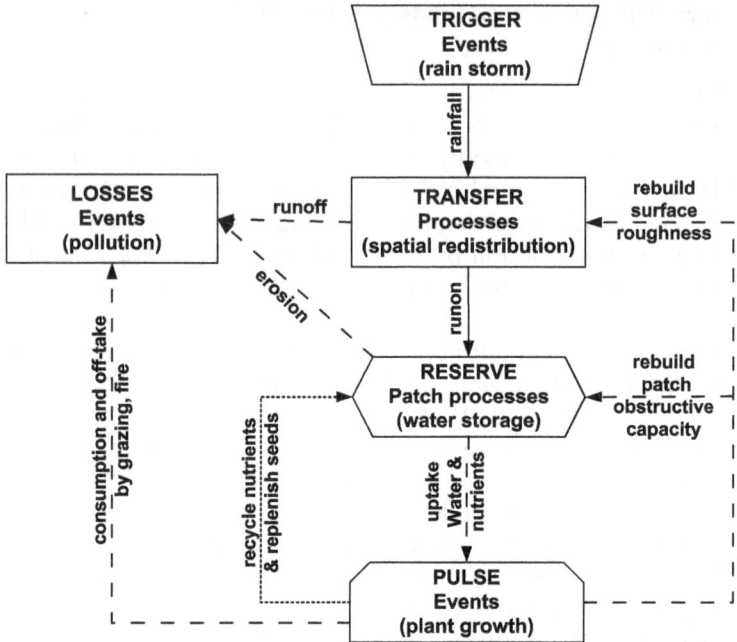

Figure 1.1 The conceptual framework representing sequences of land-scape ecological and hydrological processes and feedback loops (modified from Ludwig et al., 2005).

(Wilcox et al., 2019). It considers the landscape as a biophysical system and illustrates how the ecohydrologic interactions take place during a rainfall event (Chamizo et al., 2016). TTRP is based on the economy of vital resources and focuses on the processes that regulate the spatial movement and use of water, topsoil, and organic matter in the landscape (Tongway and Hindley, 2004).

As illustrated in Figure 1.1, the hillside receives water (*representing the trigger in the framework*) in the form of rainfall which is to be relocated across the hillslopes either by natural or man-made mechanisms. The trigger (water and/or resources) may be transferred by either getting lost through runoff from the system (e.g. erosion) or absorbed in a reserve (kept as soil moisture). The reserve is then used to create a pulse, such as new growth of vegetation or the vegetation may be kept in the reserve. With the growth of plants, some seedlings may die and be lost from the system due to herbivory or fire and the rest of the vegetation is recycled into the reserve of the system. Then,

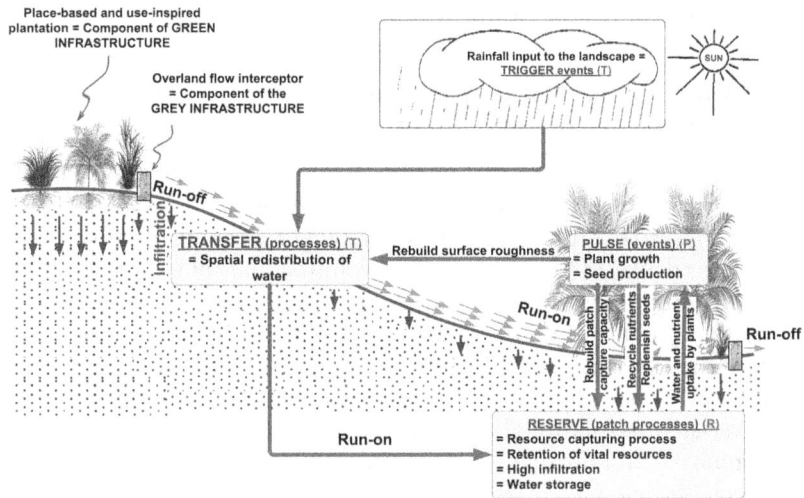

Figure 1.2 Conceptual diagram to illustrate how the green- (semi-) gray infrastructure is facilitating the functionality of a given landscape system in the context of TTRP framework.

the pulse may give resources back to the system, such as dead plant materials which serve as nutrients. The more functional a landscape is, the less resources will be lost from the system. The framework encapsulates ecosystem functional processes by assessing the dynamic spatial distribution of vital resources/natural capitals. Losses of materials are natural processes, but it is problematic when 'conserving' systems become 'leaky' via anthropogenic disturbance and losses exceed gains (Belnap et al., 2005). It interprets plant response in terms of the interaction of climatic events and soil properties, and recognizes the importance of both biotic and abiotic feedback processes that stabilize the overall system behavior.

According to the landscape function theory, restoration or replacement of missing or ineffective processes in the landscape will improve the functionality of a landscape. This perception of landscape restoration enables to make a difference in conserving the remaining forests, and re-establish forest cover in deforested and highly altered forest areas, in the mixed-use landscapes such as the tropics and sub-tropics (Chazdon and Guariguata, 2018). Technically, the basic aim of the proposed EcoLaR approach is to assist a given landscape to perform its ecohydrologic functions. The subsequent chapter also demonstrates the application of this logical framework to other types of land uses such as farm lands, gullies, and buffer zones.

1.5 Water flow regulation as a framework for effective landscape restoration: EH principle 1

Water is a strategic resource as well as a limiting and driving factor of natural processes and civilization development (Jarosiewicz et al., 2021). It is also the fundamental component of all living organisms and the major driver of biogeochemical evolution and thus of biodiversity and bio-productivity (Zalewski, 2014). Jørgensen (2016) noted that consciously controlling the hydrological retention time is a prerequisite for the application of EH in environmental management, for water is the key driver of ecosystem dynamics and the major determinant of ecosystem structure (Zalewski, 2021a). Water also drives carbon, phosphorus, and nitrogen cycling and determines ecosystem services for society (Zalewski, 2021b).

Water cycling is a key element of all environmental processes (Zalewski, 2002), but it is only useful for human activities if it is regulated in terms of time of availability, proper location, and satisfactory quality. If not, it is likely to be a burden rather than a resource (Koudstaal et al., 1992) for the landscape in general. Water flow regulation is the first critical juncture of the water cycle on the soil surface, where water either infiltrates or becomes overland flow (Wilcox et al., 2017). It is evident that water-flux regulation favors regulation of ecosystem processes that in turn is translated into regulating ecosystem services (MEA, 2005) at a landscape scale. This water flow regulation satisfied the hierarchy of factors which signify the necessity of regulating the abiotic factor (hydrology) in order to initiate regulatory feedbacks from biotic responses. The expected ecohydrologic outcomes as a result of regulation of the water cycling includes: water purification, soil erosion control, flood protection, climate regulation, and regulation of frequency and intensity of natural hazards' flow (Kandziora et al., 2013). In this sense, Zalewski (2002) defines EH as the impact of hydrology on ecosystems and vice versa.

1.6 Regulated water flow for green feedbacks and other ecological processes as a target: EH principle 2

Biodiversity, which is a fundamental indicator of human wellbeing and the prospects for sustainable future (MEA, 2005), is driven by water availability that in turn determines plant yield (Eamus et al.,

2006). In this context, the role of plant–water relations is of central interest and major focus of EH because plants occupy a key component of the hydrologic cycle (Asbjornsen et al., 2011). On the one hand, we know that plants need water to survive, and thus, the distribution, composition, and structure of plant communities are directly influenced by spatiotemporal patterns in water availability. Plants are a primary conduit for returning terrestrial water to the atmosphere (Chapin et al., 2002) while mediating albedo and roughness (Pielke et al., 1998), thereby exerting a strong effect on hydrologic fluxes of the terrestrial-atmospheric system. The EcoLaR approach consciously involves place-based and use-inspired plantation practices (both trees and non-trees) in the landscape for their multiple ecosystem services especially for their green-feedback to regulate the hydrology. Meanwhile, resilience is to be built by transforming the multiple ecosystem services in landscapes from degraded to intermediate or beyond, through manipulation of water flows or water characteristics (Falkenmark, 2016). By doing this, we are regulating (stabilizing) the dominant abiotic processes (hydrology) that in turn facilitate biotic interactions start to manifest themselves (Zalewski and Naiman, 1985).

1.7 Green- (semi-) gray infrastructure as a methodology for dual regulation: EH principle 3

The novelty of EH lies on not only understanding the complexity of water-biota interplay but also developing a methodology of how to use the ecosystem properties and the processes as a management tool, often complementary to other water resources management measures (Zalewski et al.,2004). One possible way to develop better solutions is integration of engineering, biotechnology, and EH (Zalewski, 2011). Combining 'green' ecosystem conservation and restoration with 'gray' conventional engineering approaches – using a hybrid green-gray approach – can generate more benefits for people and nature than either strategy applied alone (GGCP, 2020).

The proposed green- (semi-) gray system, which is the key component of the EcoLaR approach, basically incorporates the essence of 'dual regulation' which represents the third principle of EH. However, in order to strengthen the dual regulation between blue and green elements of the infrastructure (ecological engineering or natural solutions), the gray infrastructure (hydrotechnical measures or the conventional solutions or conventional engineering) may serve as an enabler in some cases such as mitigation against climate-driven extreme events. In the EcoLaR approach, the gray infrastructure is introduced by extrapolating the ABRC (Zalewski and Naiman, 1985) into the terrestrial phase of EH that notify once water flow is stabilized, the biotic feedbacks become dominant and favors higher biodiversity. The role of the gray component of the proposed infrastructure is to assist nature to do its job. In the context of EcoLaR approach, the physical barriers that restrict overland flow are meant to provide necessary conditions of enhancing the biota (*the second ecohydrologic principle*) and the subsequent dual regulation once the vegetation, which occupies a key component of the hydrologic cycle (Asbjornsen et al., 2011), develops.

To avoid over-engineering of the landscape, the EcoLaR approach adopted the ecological engineering design principles proposed by Bergen et al. (2001) with little modification (*the last three principles are newly added by the author of this book*) as follows:

1 Design consistent with ecological principles = *treat nature as a partner in design*
2 Design for site-specific context = *solutions should be site-specific and small-scale*
3 Maintain the independence of design functional requirements = *Best designs are those that have independent (not coupled) functional requirements*
4 Design for efficiency in energy and information = *let nature do some of the engineering*
5 Acknowledge the values and purposes that motivate design = *design practices that acknowledge the motivating values and purposes will be more successful*
6 Encourage minimum earth work
7 Use local materials and capacity as much as possible
8 Acknowledge indigenous knowledge

1.8 Understanding the underpinning limitation of the current practices of landscape management

The first step in introducing the ecohydrologic strategies into the current practices of landscape restoration is to understand the underpinning limitations of those practices. The subsequent sub-sections discuss the top ten limitations of the conventional practice of environmental management as majorly sourced from Belete (2021):

1.8.1 Limitation 1: underutilized opportunity of dual regulation between 'biota and hydrology'

According to decision theory (Figure 1.3), every strategy for success including sustainable water resources management has to contain both elimination of threats and amplification of opportunities (Zalewski, 2006). In relation to water resources management challenges, the major threats can be grouped as: (1) acceleration of the water and matter circulation due to degradation of natural plant communities; (2) emission of large fluxes of mineral, organic, and chemical substances to the environment; and (3) reduction of habitats for plants and animals in terms of space and connectivity (Zalewski, 2014). Meanwhile, the modification of the landscape and the consequent land cover-driven hydrological cycle went so far that it is impossible to eliminate the ever intensively occurring threats – water deficits and floods – with hydrotechnical measures alone, calling for an urgent new approach for sustainability (Zalewski, 2014).

As indicated in Figure 1.3, the mechanistic approach efforts to compensate the prevailing degradation of water resources and environment (*bottom line of the figure*) have not been successful (Kundzewicz, 1999), and they over-engineer the environment (Zalewski, 2002). This

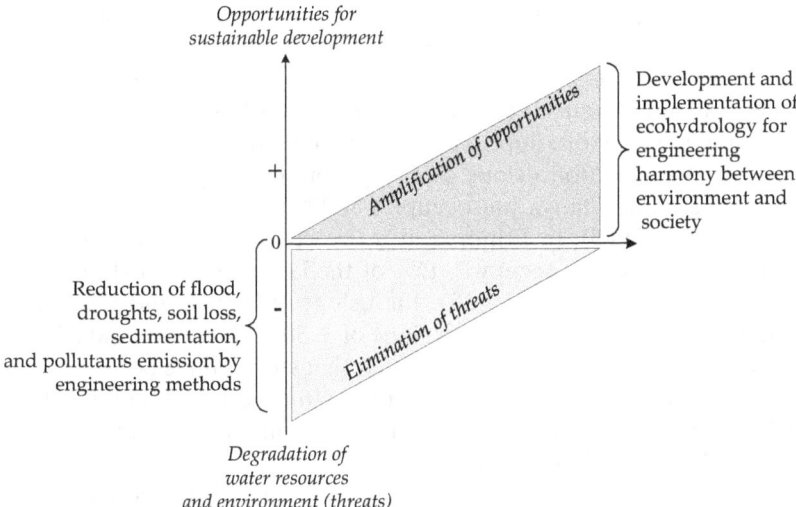

Figure 1.3 The basic theory of decision (modified after Zalewski, 2004).

limitation is likely emanated from the fact that water resources at the basin scale are the result not only of climatic conditions and geomorphologic structures, but also, to a great extent, of biological evolution and succession (Zalewski, 2002). In this regard, our conventional practices can likely be concluded as a half-way journey to sustainability.

1.8.2 Limitation 2: significant size of scarce productive lands are occupied and becoming out of production due to the physical land management technologies

Land degradation is a major cause of poverty in rural areas of sub-Saharan Africa (Mesfin et al., 2018) and Africa in general where 50% of the total erosion-affected people are concentrated (FAO, 2002). It has caused reduction of soil fertility and crop yield (Amare et al., 2013), challenging the economic and social wellbeing of the current and future generations (Haregeweyn et al., 2012; Keno and Suryabhagavan, 2014). For instance, in order to curb the effects of land degradation in Ethiopia, the government has been taking serious measures (Wordofa et al., 2020) such as expanding soil and water conservation practices throughout the country (Adimassu et al., 2012). Despite this effort, farmers were not enthusiastic enough in widely accepting the technologies (Wood, 1990).

The commonly mentioned reasons for this low adoption of the technologies are linked to the socioeconomic and governance issues of the planning and implementation of the interventions (Asfaw and Neka, 2017; Asnake et al., 2018; Kirubel and Gebreyesus, 2011). However, the technical design issue of the physical structures has been significantly affecting the level of adoption of the technologies. Paradoxically, depending on slope (for a slope category from five to greater than 55%) and soil stability, fanya juu occupies 8–40% of cultivable land areas (Tenge et al., 2005). In Ethiopia, it was recommended that fanya juu terraces (Figure 1.4) occupy 2–15% of the land area for a slope of 3–15%; stone bunds occupy 5–25% for a slope of 5–50%; and soil bunds (Figure 1.5) occupy 2–20% for a slope of 3–30% (Teshome et al., 2013). Vancampenhout et al. (2006) estimated that stone bunds occupy about 8% of the farmland in northern Ethiopia. In experimental plots established in the central highlands of Ethiopia, soil bunds occupy 8.6% of cultivable land (Adimassu et al., 2012).

In confirming the paradoxical impact of some of the conventional land management practices, a recent report of Biratu et al. (2021) showed that soil bund and fanya juu reduced the grain yield by 24 and 22%, respectively, on average, and they reduced off-site erosion at

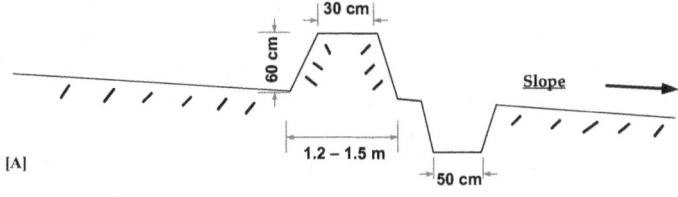

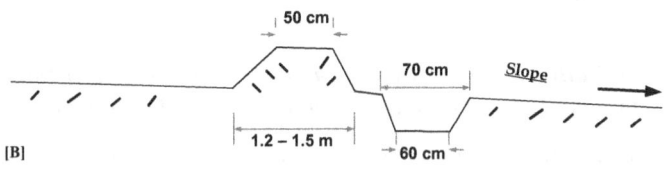

Figure 1.4 The basic technical specification of fanya juu in the conventional system for (a) stable and (b) unstable soils (modified from: Daniel, 2001; Desta et al., 2005; Hurni et al., 2016; Mitiku et al., 2006).

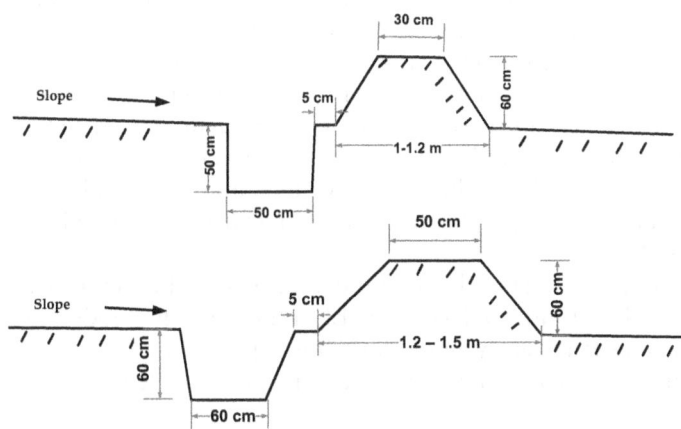

Figure 1.5 The basic technical specification of soil bunds in the conventional system for (a) stable and (b) unstable soils (modified from: Daniel, 2001; Desta et al., 2005; Hurni et al., 2016; Mitiku et al., 2006).

the expense of poor farmers who can ill-afford any additional costs. Farmers are very curious about the yield effect of the technology since the structures take up productive land and maintenance is often labor-intensive and costly (Shiferaw and Holden, 1998). In this line, it is evident that farmers with larger farm size can afford retaining structures compared to those with relatively lower farm size (Birhanu and Meseret, 2013). To truly achieve such 'win-win' outcomes, much more attention to the interaction between SWC technologies and production factors, such as land, labor, and weather endowments, is needed (Kassie et al., 2008).

1.8.3   Limitation 3: the existing technologies over-engineer the environment and become less appealing to the community

The conventional hydrotechnical solutions such as masonry retaining walls, gabion wire mesh, and check dams (Figure 1.6) have been constructed along and across the gully network with little success, high cost, and addressing only short-term and single goals (Naiman et al., 1995). In the eyes of sustainability, it is apparently observed that it is impossible to eliminate the ever intensively occurring threats such as floods (Ryszkowski and Kedziora, 1999) with hydrotechnical measures alone (Zalewski, 2014). In many situations, this mechanistic approach has led to over-engineering of the environment which seriously reduces the role of ecological processes in moderating the water cycle (Zalewski, 2002). It is now apparent that engineers, who have actually become real decision makers in environmental management (Zalewski, 2014), shall understand that the water resources at the basin scale are the result not only of climatic conditions and geomorphology structures but also, to a great extent, of biological evolution and succession (Zalewski, 2002).

With this understanding in mind, all range of the engineering methods should be harmonized with EH regulation methods toward the enhancement of ecosystem's carrying capacity, and compensate the cumulative human impact on the environment (Zalewski, 2014).

1.8.4   Limitation 4: water and nutrient cycle regulation is rarely targeted in the conventional land management practices

Water should be a strategic resource in watershed management. It is the primary factor of sustainable development, eradication of poverty, and reversal of ecosystem degradation (Zalewski, 2010) that limits and regulates the ability of ecosystems to accumulate carbon,

Figure 1.6 A failed gabion wire mesh implying the likely caused by incompatibility of the soil nature with the massive weight of the structure.
Photograph by the author.

nitrogen, and phosphorus (Zalewski, 2015); the most important natural resource (Tidwell, 2016) and agent of all ecological processes (Zalewski et al., 2003); the most fundamental driver of ecological processes (Chapin et al., 2002); and the essential reactant, catalyst, or medium for many biogeochemical reactions (Wang et al., 2015); the primary factor to reverse ecosystem degradation; and major driver of biogeochemical evolution, and hence, of biodiversity and bioproductivity (Zalewski, 2015); a central organizer of ecosystems for eradication of poverty (Tidwell, 2016; Zalewski, 2010); a key aspect of sustainability challenges of humans in the Anthropocene (Sivapalan et al., 2014); the only factor linking all ecosystem services (Krauze and Wagner, 2008); the most critical factor to achieve sustainable development (Zalewski, 2002); and fundamental for all forms of life and most human activities on Earth (Zalewski, 2015). It also works as a carrier of solutes, plays a key role in local climate regulation, and sets the ecohydrologic conditions for biological diversity in any habitat (Rockström et al., 1999).

1.8.5 Limitation 5: the biological measures are considered as stabilizers of the physical measures (not as dual regulators as in the case of ecohydrology)

In most cases of the conventional watershed management system, biological measures of soil and water conservation are thought to enhance efficiency of soil erosion protection by stabilizing the physical structures (Amare et al., 2014). However, in addition to the above role, the biota has significant opportunity to interact with hydrology for the ecohydrologic effect of dual regulation.

1.8.6 Limitation 6: land management practices shall derive multiple ecosystem services (ES) as outcomes for better adoption by farmers

Land management practices need to reflect today's diverse goals, which increasingly include the recovery and improvement of landscape function to support multiple ecosystem services (Aronson et al., 2006). Ecosystem services, the benefits that humans obtain from ecosystems, are vital for rural livelihoods (Bhatta et al., 2015). One of the underlying assumptions for involving communities in natural resource management is the fact that people conserve a resource only if benefits exceed the costs of conservation, and also people conserve a resource that is directly adding to their quality of life (Thakadu, 2005).

In practice, ecohydrologic processes are closely connected to many water-related ecosystem services including climate moderation, water supply and quality, and flood mitigation (Ellison et al., 2012; Jackson et al., 2009; Sun et al., 2015). Ecosystem services that are potentially derived from restoration efforts are those benefits that are directly enjoyed, consumed, or used to yield human 'wellbeing' (Boyd and Banzhaf, 2006) and can be either intermediate or final services (Tallis and Polasky, 2011). For instance, intermediate ecosystem services can be the ecological and hydrological processes and the final services refer to aspects of the environment that have direct value to society (Ringold et al., 2013). The ecosystem services approach refers to the interdependencies between nature and human wellbeing (Schleyer et al., 2017; Steger et al., 2018). They are increasingly used worldwide as a framework for the purposes of ecological restoration and conservation (Wei et al., 2017), watershed management (Falkenmark et al., 2004), and sustainable development policymaking (Asbjornsen et al., 2015).

One of the key sustainability dimensions of the multiple ecosystem services is the contribution to community livelihood. Chamber and Conway (1992) describe livelihood as a system comprising of assets, capabilities, and activities for means of living. Ecosystem degradation is a major threat to livelihoods and sustainable development across the globe (Battistelli et al., 2021), and the quality of people's life is directly depending on how well the vital resources in the landscape are conserved (Matondo, 2002; Worku et al., 2018). Given the mounting challenges of poverty, low economic development and poor agricultural system, the issue of livelihood sustainability is a growing concern especially in developing countries (Ashley, 2000; Baumann, 2000). In this regard, landscape restoration can address poverty eradication in two ways (Shixiong, 2017). First, restoration activities generate employments, thus improving the socioeconomic conditions of the poor (Abhilash et al., 2016). Second, the restored land supports increased future production as well as improved ecosystem services. To realize this truth, the course of actions in the landscape restoration is required to achieve one of its social objectives.

1.8.7 Limitation 7: the importance of gaining immediate benefits for the farmers

The subsistence farmer cannot afford to respond to philosophical or emotional appeals to care for the soil, and this means that conservation measures must have visible short-term benefits to the farmer (Tilahun and Belay, 2019). Dialla (1992) acknowledged that third world farmers are very responsive to immediate observable outcomes rather than to uncertain, long-term benefits. In this action research, the ecological engineering interventions are meant to provide multiple ecosystem services among which some deliver immediate benefits to the farmers (e.g., grasses for cattle; protecting the community from flood)

1.8.8 Limitation 8: highly degraded and dysfunctioning landscapes are perceived to slowly rehabilitate themselves

Conceptual confusion seems to revolve around management of degraded and dysfunctioning landscapes. Some of these landscapes might have crossed the state where their capacity for regeneration is greatly reduced or lost, recovery is arrested, core interactions and feedbacks are broken, and human intervention is required to initiate a trajectory of recovery (Ghazoul and Chazdon, 2017).

1.8.9 Limitation 9: necessity of climate-smart landscape management approach

Climate change is a new reality to the conventional watershed management practices and recognized as one of the major derivers of ecosystem change (MEA, 2005). Its impacts are now inevitable and are expected to affect people in African countries the most (Joyce et al., 2006). Moreover, it has become a major concern to the sustainability of water resources and ecosystems (Houghton et al., 1996).

As a component of this big picture, some initiatives such as climate-smart agriculture target re-orientation of the conventional system to support food security under the new realities of climate change (Lipper et al., 2014). By the same token, the conventional watershed management practices need to customize themselves to the ever-strengthening climate-related phenomena like extreme flood and drought. By climate-smart landscape, we are referring to a landscape that simultaneously supports climate, development, and conservation objectives (Kusters, 2015) which has emerged parallel to the development of climate-smart discourse within the agricultural development and conservation communities (Scherr et al., 2012). Natural resource-based sustainability refers to the ability of a system to maintain productivity when subjected to disturbing forces, stresses, and shocks (Joyce et al., 2006). As indicated in the multi-functionality of the proposed system (*reason 6*), climate change regulation is one of the ecosystem services to be offered. By restoring the landscape, we are providing two key management approaches to mitigate the effects of climate change: sequestering carbon through the establishment of green biomass; and the conservation and restoration of biodiversity and ecosystem services (SER, 2019). Lindenmayer et al. (2002) reported that landscape-scale restoration approaches are likely key for arresting and mitigating negative effects of climate change. Ecohydrologic processes are also closely connected to many water-related ecosystem services including climate moderation, water supply and quality, and flood mitigation (Ellison et al., 2012; Jackson et al., 2008; Sun et al., 2015).

1.8.10 Limitation 10: necessity of systemic approach

A systems approach, which refers to arrangement of things to form a whole, must be used to identify and solve interrelated watershed problems for it has a number of interrelated problems that require an integrated solution. This approach provides a process-based framework to define watershed and channel dynamics, and to develop integrated solutions (Watson et al., 1999). However, little effort is put into developing new methods and systemic solutions for sustainable environmental management (Zalewski, 2011). Hence, a paradigm shift to process-oriented thinking in entire human activity is necessary for effective implementation of system solutions in the holistically perceived catchment environment (Zalewski, 2014).

1.9 Conclusion and the way forward with ecohydrology-based landscape restoration (EcoLaR)

It is observed that the contemporary landscape management practices in developing countries are not sufficient to guarantee sustainable development, and they need to be renovated. Despite tangible benefits of the conventional technologies of landscape management practices, many ecosystem services are adversely affected when the ecosystems are degraded due to construction and their tendency to over-engineer the ecosystem. These gaps potentially create broken or loose connections along the result chain of the road to sustainability. The essence of recognizing the prevailing gaps in land management practices in developing countries is to devise a better strategy so as to align with the UN sustainable development goals. The proposed approach of EcoLaR spins off from these problems and strives to fine-tune those limitations toward effective and sustainable management of natural resources through its theoretical as well as practical experimentation.

References

Abhilash, P.C., Tripathi, V., Edrisi, S.A., Dubey, R.K., Bakshi, M., Dubey, P.K., Singh, H.B., & Ebbs, S.D. (2016). Sustainability of crop production from polluted lands. *Energy, Ecology and Environment, 1*(1), 54–65.

Adimassu, Z., Kessler, A., & Hengsdijk, H. (2012). Exploring determinants of farmers' investments in land management in the Central Rift Valley of Ethiopia. *Applied Geography, 35*, 191–198.

Agostini, P., & Purdie, E. (2017). *Landscape degradation: A world of landscape restoration opportunities.* World Bank data blogs. *https://blogs. worldbank.org/opendata/landscape-degradation-world-landscape-restoration-opportunities*

Amare, T., Terefe, A., Selassie, Y.G., Yitaferu, B., Wolfgramm, B., & Hurni, H. (2013). Soil properties and crop yields along the terraces and toposequence of Anjeni Watershed, central highlands of Ethiopia. *Journal of Agricultural Sciences, 5*, 134–144.

Amare, T., Zegeye, A.D., Yitaferu, B., Steenhuis, T.S., Hurni, H., & Zeleke, G. (2014). Combined effect of soil bund with biological soil and waterconservation measures in the northwestern Ethiopian highlands. *Ecohydrology and Hydrobiology, 14*, 192–199.

Amede, T. (2003). *Restoring soil fertility in the highlands of east Africa through participatory research.* International Center for Research in Agroforestry. AHI brief No. A1.

Ariti, T., van Vliet, J., & Verburg, P. (2015). Land-use and land-cover changes in the Central Rift Valley of Ethiopia: Assessment of perception and adaptation of stakeholders. *Applied Geography, 65*, 28–37.

Aronson, J., & Alexander, S. (2013). Ecosystem restoration is now a global priority: Time to roll up our sleeves. *Restoration Ecology, 21*(3), 293–296.

Aronson, J., Clewell, A.F., Blignaut, J.N., & Milton, S.J. (2006). Ecological restoration: A new frontier for nature conservation and economics. *Journal for Nature Conservation, 14*, 135–139.

Asbjornsen, H., Goldsmith, G.R., Alvarado-Barrientos, M.S., Rebel, K., van Osch, F.P., Rietkerk, M., & Dawson, T.E. (2011). Ecohydrological advances and applications in plant-water relations research: A review. *Journal of Plant Ecology, 4*(1–2), 3–22.

Asbjornsen, H., Mayer, A.S., Jones, K.W., Selfa, T., Saenz, L., Kolka, R.K., & Halvorsen, K.E. (2015). Assessing impacts of payments for Catchment services on sustainability in coupled human and natural systems. *Bioscience, 65*(6), 579–591.

Asfaw, D., & Neka, M. (2017). Factors affecting adoption of soil and water conservation practices: The case of Wereillu Woreda (District), South Wollo Zone, Amhara Region, Ethiopia. *International Soil and Water Conservation Research, 5*(4), 273–279.

Ashley, C. (2000). *Applying livelihood approaches to natural resource management initiatives: Experiences in Namibia and Kenya* (Working Paper 134). London: ODI.

Asnake, M., Heinimann, A., Gete, Z., & Hurni, H. (2018). Factors affecting the adoption of physical SWC practices in the Ethiopian highlands. *International Soil and Water Conservation Research, 6*(1), 23–30.

Baig, S, Rizvi, A.R., & Jones, M. (2017). *Enhancing resilience through forest landscape restoration: Understanding synergies and identifying opportunities* (Discussion Paper). Gland: IUCN.

Battistelli, F., Tarekegn, M., & Shiferaw, M. (2021). *Restoring Ethiopia's ecosystems can support livelihoods and COVID-19 recovery.* Commentary. World Resources Institute. *https://www.wri.org/insights/restoring-ethiopias-ecosystems-can-support-livelihoods-and-covid-19-recovery*

Baumann, P. (2000). *Sustainable livelihoods and political capital: Arguments and evidence from decentralization and natural resource management in India* (Working Paper 136). London: Overseas Development Institute.

Belete, M.D. (2021). Review of the underpinning reasons and field demonstrations to incorporate ecohydrologic strategy into landscape restoration in water-limited ecosystems. *Ecohydrology and Hydrobiology, 21*(3), 529–542.

Belnap, J., Welter, J., Grimm, N., Barger, N., & Ludwig, J. (2005). Linkages between microbial and hydrologic processes in arid and semiarid watersheds. *Ecology, 86*, 298–307.

Bergen, S.D., Bolton, S.M., & Fridley, J. (2001). Design principles for ecological engineering. *Ecological Engineering, 18*(2), 201–210.

Bhatta, L.D., van Oort, H., Stork, N.E., & Baral, H. (2015). Ecosystem services and livelihoods in a changing climate: Understanding local adaptations in the Upper Koshi, Nepal. *International Journal of Biodiversity Science, Ecosystem Services & Management, 11*(2), 145–155.

Biratu, A.A., Bedadi, B., Gebrehiwot, S.G. Hordofa, T., Asmamaw, D.K, & Melesse, A.M. (2021). Implications of land management practices on selected ecosystem services in the agricultural landscapes of Ethiopia: A review. *International Journal of River Basin Management, 2021*(1), 1–18.

Birhanu, A., & Meseret, D. (2013). Structural soil and water conservation practices in Farta District, North Western Ethiopia: An investigation on factors influencing continued use. *Science, Technology and Arts Research Journal, 2*(4), 114–121.

Boyd, J., & Banzhaf, S. (2006). *What are ecosystem services? The need for standardized environmental accounting units* (Resource for the Future Discussion Paper 06-02).

Boyd, J., & Banzhaf, S. (2007). What are ecosystem services? The need for standardized environmental accounting units. *Ecological Economics, 63*, 616–626.

Brasser, A., & Ferwerda, W. (2015). *4 Returns from landscape restoration. A systemic and practical approach to restore degraded landscapes.* Amsterdam: Commonland Publication.

Bunch, M.J., Parkes, M., Zubrycki, K., Venema, H., & Hallstrom, L. (2014). Watershed management and public health: An exploration of the intersection of two fields as reported in the literature from 2000 to 2010. *Environmental Management, 54*, 240–254.

Burdyuzha, V. (2006). *The future of life and the future of our civilization.* Dodrecht: Springer.

CBD (2004). *The ecosystem approach.* CBD guidelines: Secretariat of the convention on biological diversity. Montreal.

CBD (2009). *Connecting biodiversity and climate change mitigation and adaptation* (Technical Series 41). Montreal, Canada: Report of the Second Ad Hoc Technical Expert Group on Biodiversity and Climate Change.

CBD (2012). *Most used definitions/descriptions of key terms related to ecosystem restoration.* Hyderabad: CBD.

Chamber, R., & Conway, G. (1992). *Sustainable rural livelihoods: Practical concepts for the 21st century* (IDS Discussion Paper 296). Brighton: Institute of Development Studies.

Chamizo, S., Cantón, Y., Rodríguez-Caballero, E., & Domingo, F. (2016). Biocrustspositively affect the soil water balance in semi arid ecosystems. *Ecohydrology, 9*, 1208–1221.

Chapin, F.S., Matson, P.A., & Mooney, H.A. (2002). *Principles of terrestrial ecosystem ecology.* New York: Springer.

Chazdon, R.L., & Guariguata, M.R. (2018). *Decision support tools for forest landscape restoration: Current status and future outlook* (Occasional Paper 183). Bogor: CIFOR.

Chen, S.J. (2007). *History of China water resource management*. Beijing: China Water Resources Management Publishing House.

Costanza, R., & Mageau, M. (1999). What is a healthy ecosystem?. *Aquatic Ecology 33*, 105–115.

D'Odorico, P., Okin, G.S., & Bestelmeyer, B.T. (2012). A synthetic review of feedbacks and drivers of shrub encroachment in arid grasslands. *Ecohydrology, 5*, 520–530.

Daniel, D. (2001). *Soil and water conservation manual/guideline for Ethiopia*. Addis Ababa: Ethiopia Soil and water Conservation Team, Natural Resources Management and Regulatory Department, Ministry of agriculture.

Deal, R.L., Cochran, B., & LaRocco, G. (2012). Bundling of ecosystem services to increase forestland value and enhance sustainable forest management. *Forest Policy and Economics, 17*, 69–76.

Descheemaeke, K., Nyssen, L., Rossi, J., Poesen, J., Haile, M., Raes, D., Muys, B., Moeyersons, J., & Deckers, S. (2006). Sediment deposition and pedogenesis in exclosures in the Tigray Highlands, Ethiopia. *Geoderma, 132*, 291–314.

Desta, G., Nyssen, J., Poesen, J., Deckers, J., Mitiku, H., Govers, G., & Moeyersons, J. (2005). Effectiveness of stone bunds in controlling soil erosion on cropland in the Tigray highlands, Northern Ethiopia. *Soil use and management, 21*, 287–297.

Dialla, B.E. (1992). *The adoption of soil and water conservation practices in Burkina Faso: The role of indigenous knowledge, social structure and institutional support*. Arbor: A Bell and Howell Information Company.

Dudley, N. (Ed.) (2008). *Guidelines for Applying Protected Area Management Categories*. Gland: IUCN.

Eamus, D., Hatton, T., Cook, P., & Colvin, Ch. (2006). *Ecohydrology-Vegetation function, water and resource management*. Collingwood: Commonwealth Scientific and Industrial Research Organization.

ELD (2015). *The value of land: Prosperous lands and positive rewards through sustainable land management*. ELD. Retrieved from www.eld-initiative.org

Ellison, D., Futter, M.N., & Bishop, K. (2012). On the forest cover-water yield debate: From demand- to supply-side thinking. *Global change biology, 18* (3), 806–820.

Falkenmark, M. (2003). Freshwater as shared between society and ecosystems: From divided approaches to integrated challenges. *Philosophical Transactions of the Royal Society of London Series B-Biological Sciences, 358*(1440), 2037–2049.

Falkenmark, M. (2016). Water and human livelihood resilience: A regional-to-global outlook. *International Journal of Water Resources Development, 33* (2).

Falkenmark, M., Gottschalk, L., Lundqvist, J., & Wouters, P. (2004). Towards integrated catchment management: Increasing the dialogue between scientists, policymakers and stakeholders. *International Journal of Water Resources Development, 20*(3), 297–309.

FAO (2002). *Land-water linkages in rural watersheds*. Rome: Land and Water Bulletin 9.

FAO (2012). *Mainstreaming climate-smart agriculture into a broader landscape approach*. Rome: Background Paper for the Second Global Conference on Agriculture, Food Security and Climate Change. Food and Agricultural Organization of the United Nations.

FAO, IUCN CEM & SER (2021). *Principles for ecosystem restoration to guide the United Nations Decade 2021–2030*. Rome: FAO.

GGCP (Green-Grey Community of Practice) (2020). Practical guide to implementing green-gray infrastructure. Friends of ecosystem-based adaptation. Retrieved from https://friendsofeba.com/wgs/green-gray/.

Ghazoul, J., & Chazdon, R. (2017). Degradation and Recovery in Changing Forest Landscapes: A Multiscale Conceptual Framework. *Annual Review of Environment and Resources, 42*(1), 161–188.

Haregeweyn, N., Berhe, A., Tsunekawa, A., Tsubo, M., & Meshesha, D.M. (2012). Integrated watershed management, an effective approach to curb land degradation: A case study of the Enabered watershed, northern Ethiopia. *Journal of Environmental Management, 50*, 1219–1233.

Haregeweyn, N., Poesen, J., Nyssen, J., De Wit, J., Haile, M., Govers, G., & Deckers, S. (2006). Reservoirs in Tigray: Characteristics and sediment deposition problems. *Land Degradation and Development, 17*, 211–230.

Houghton, J.T., Meira Filho, L.G., Callander, B.A., Harris, N, Kattenberg, A., Maskell, K., & Lakeman, J.A. (Eds). (1996). *Climate change 1995: The science of climate change*. Great Cambridge: UN digital library. Cambridge University Press, for the Intergovernmental Panel on Climate Change.

Hurni, H., Berhe, W.A., Chadhokar, P., Daniel, D., Gete, Z., Grunder, M., & Kassaye, G. (2016). *Soil and water conservation in Ethiopia: Guidelines for development agents*. Bern: Centre for Development and Environment (CDE); Bern, Open Publishing (BOP).

IHP (2013). *Ecohydrology for Sustainability*. UNESCO.

IUCN & WRI (2014). *A guide to the restoration opportunities assessment methodology (ROAM):assessing forest landscape restoration opportunities at the national or sub-national level* (Working Paper [Road-test Edition]). Gland: IUCN.

IUCN (2017). *The IUCN red list of threatened species (V2017-3)*. Gland: IUCN.

Jackson, E.C., Krogh, S.N., & Whitford, W.G (2009). Ecohydrology bearings-invited commentary. Ecohydrology in a human-dominated landscape. *Ecohydrology, 2*, 383–389.

Jackson, R.B., Randerson, J.T., Canadell, J.G., Anderson, R.G., Avissar, R., Baldocchi, D.D., Bonan, G.B., Caldeira, K., Diffenbaugh, N.S., & Field, C.B. (2008). Protecting climate with forests. *Environtal Research Letters, 3*(4).

Jarosiewicz, P., Jurczak, J., & Zalewski, M. (2021). *Ecohydrology for sustainable urban water management*. Pre-Conference for the Second International Conference on Water, Megacities and Global Change. Paris: UNESC.

Jørgensen, S.E. (2016). Ecohydrology as an important concept and tool in environmental management. *Ecohydrology & Hydrobiology, 16*(1), 4–6.

Joyce, L., Haynes, R., White, R., Barbour, R., & James, R. (2006). *Bringing climate change into natural resource management: Proceedings* (Gen. Tech. Rep. PNW-GTR-706). Portland, OR: U.S. Department of Agriculture, Forest Service, Pacific Northwest Research Station.

Kandziora, M., Burkhard, B., & Müller, F. (2013). Interactions of ecosystem properties, ecosystem integrity and ecosystem service indicators-a theoretical matrix exercise. *Ecological Indicators, 28*, 54–78.

Kassie, M., Holden, S., Köhlin, G., & Bluffstone, R. (2008). *Economics of soil conservation adoption in high-rainfall areas of the Ethiopian highlands.* Environment for Development (Discussion Paper Series: EfD DP 08–09). Sweden: Environment for Development.

Kebede, W. (2015). Evaluating watershed management activities of campaign work in Southern Nations, Nationalities and Peoples' Regional State of Ethiopia. *Environmental Systems Research, 4* (6).

Keno, K., & Suryabhagavan, K.V. (2014). Multi-temporal remote sensing of landscape dynamics and pattern change in Dire district, Southern Ethiopia. *Journal of Geomatics, 8*(2), 189–194.

Khan, N.H., Nafees, M., Rahman, A., & Saeed, T. (2021). Ecodesigning for ecological sustainability. In T. Aftab, K. R. Hakeem (Eds.), frontiers in plant-soil interaction-molecular insights into plant adaptation. Cambridge, Massachusetts: Academic Press.

Kirubel, M., & Gebreyesus, B. (2011). Impact assessment of soil and water conservation measures at medego watershed in Tigray, Northern Ethiopia. *Maejo International Journal of Science and Technology, 5*(3), 312–330.

Koudstaal, R., Rijsberman, F.R., & Savenije, H. (1992). *Water and sustainable development* (0165–0203/92,040277-14). United Kingdom: Butterworth-Heinemann Ltd.

Krauze, K., & Wagner, I. (2008). *An ecohydrologicalapproach for the protection and enhancement ofecosystem services.* In Petrosillo, I., Jones, B., Muller, F., Zurlini, G., Krauze, K., Victorov, S. (Eds।)., Use of landscape sciences for the assessment of environmental security. Springer-Verlag Publishers.

Kundzewicz, Z.W. (1999). Flood protection-sustainability issues. *Hydrological Sciences, 44*(4), 559–571.

Kusters, K. (2015). *Climate-smart landscapes and the landscape approach: An exploration of the concepts and their practical implications.* Wageningen: Tropenbos International.

Lindenmayer, D.B., Manning, A.D., Smith, P.L., Possingham, H.P., Fischer. J., Oliver, I., & McCarthy, M.A. (2002). The focal-species approach and landscape restoration: A critique. *Conservation Biology, 16*(2), 338–345.

Lipper, L., Lipper, L., Thornton, P., Campbell, B.M., Baedeker, T., Braimoh, A., Bwalya, M., Caron, P., Cattaneo, A., Garrity, D., Henry, K., Hottle, R., Jackson, L., Jarvis, A., Kossam, F., Mann, W., McCarthy, N., Meybeck, A., Neufeldt, H., Remington, T., Sen, P.T., Sessa, R., Shula, R., Tibu, A., & Torquebiau, E.F. (2014). Climate-smart agriculture for food security, *Nature Climate Change, 4*(12), 1068–1072.

Ludwig, J.A., Wilcox, B.P., Breshears, D.D., Tongway, D.J., & Imeson, A.C. (2005). Vegetation patches and runoff-erosion as interacting ecohydrological processes in semiarid landscapes. *Ecology, 86*, 288–297.

Ludwig, J., Tongway, D., Freudenberger, D., Noble, J., & Hodgkinson, K. (1997). *Landscape ecology, function and management: Principles from Australia''s Rangelands*. Melbourne: CSIRO Publishing.

Machar, I. (2020). Sustainable landscape management and planning. *Sustainability, 12*, 2354.

Mansourian, S. (2005). Overview of forest restoration strategies and terms. In S. Mansourian, D. Vallauri, & N. Dudley (Eds.), *Forest restoration in landscapes: Beyond planting trees*. New York: Springer.

Matondo, J.I. (2002). A comparison between conventional and integrated water resources planning and management. *Physics and Chemistry of the Earth, 27*(11), 831–838.

MEA (Millennium Ecosystem Assessment) (2005). *Ecosystems and human well-being: Synthesis*. Washington, DC: Island Press.

Mekuria, W., Veldkamp, E., Mitiku, H., Nyssen, J., Muys, B., & Kindeya, G. (2006). Effectiveness of enclosures to restore degraded soils as a result of overgrazing in Tigray, Ethiopia. *Journal of Arid Environment, 69*, 270–284.

Mesfin, S., Taye, G., Desta, Y., Sibhatu, B., Muruts, H., & Mohammedbrhan, M. (2018). Short-term effects of bench terraces on selected soil physical and chemical properties: Landscape improvement for hillside farming in semi-arid areas of Northern Ethiopia. *Environmental Earth Sciences, 77* (11).

Mitiku, H., Herweg, K., & Stillhardt, B. (2006). *Sustainable land management – A new approach to soil and water conservation in Ethiopia*. Mekelle: Land Resources Management and Environmental Protection Department, Mekelle University; Bern: Centre for Development and Environment (CDE), University of Bern, and Swiss National Centre of Competence in Research (NCCR) North-South 269.

Naiman, R.J., Magnuson, J.J McKnight, D. M., & Stanford, J.A. (1995). *The Freshwater Imperative*. Washington: A Research Agenda, Island Press.

Nuttle, W.K. (2002). Is ecohydrology one idea or many? *Hydrological Sciences Journal, 47*, 805–807.

Osman, M., & Sauerborn, P. (2001). Soil and water conservation in Ethiopia experiences and lessons. *Soils & Sediments, 1*(2), 117–123.

Pfund, J.L., & Stadtmüller, T. (2005). *Forest landscape restoration (FLR)*. *InfoResources Focus, 2*, 1–12.

Pielke, R.A., Avissar, R., Raupach, M., Dolman, A.J., Zeng, X.B., & Denning, A.S. (1998) Interactions between the atmosphere and terrestrial ecosystems: Influence on weather and climate. *Global Change Biology, 4*, 461–475.

Porporato, A., & Rodriguez-Iturbe, I. (2013). Ecohydrology bearings-invited commentary from random variability to ordered structures: A search for general synthesis in ecohydrology. *Ecohydrology, 6*(3), 333–342.

Reed, J., Deakin, L., & Sunderland, T. (2015). What are 'Integrated Landscape Approaches' and how effectively have they been implemented in the tropics: A systematic map protocol. *Environmental Evidence, 4* (2).

Ringold, P.L., Boyd, J., Landers, D., & Weber, M. (2013). What data should we collect? A framework for identifying indicators of ecosystem contributions to human well-being. *Frontiers in Ecology and Environment, 11*(2), 98–105.

Rockström, J., Gordon, L., Folke, C., Falkenmark, M., & Engwall, M. (1999). Linkages among water vapor flows, food production, and terrestrial ecosystem services. *Conservation Ecology, 3*(2), 5.

Rodriguez-Iturbe, I. (2000). Ecohydrology: A hydrologic perspective of climate-soil-vegetation dynamics. *Water Resources Research, 36*(1), 3–9.

Ryszkowski, L., & Kedziora, A. (1999). State of the art in the appraisal of global climate change phenomena. *Geographia Polonica, 72*(2), 5–8.

Sayer, J. (2009). Reconciling conservation and development: Are landscapes the answer? *Biotropica, 41*, 649–652.

Scherr, S.J., Shames, S., & Friedman, R. (2012). From climate-smart agriculture to climate-smart landscapes. *Agriculture & Food Security, 1* (12).

Schleyer, C., Lux, A., Mehring, M., & Görg, C. (2017). Ecosystem services as a boundary concept: Arguments from social ecology. *Sustainability, 9* (1107).

SER (Society for Ecological Restoration International Science) & PWG (Policy Working Group) (2004). *The SER international primer on ecological restoration.* Arizona, USA: Society for Ecological Restoration.

SER (Society for Ecological Restoration) (2008). *Ecological Restoration as a tool for reversing ecosystem fragmentation* (Policy position statement). Arizona, USA: SER.

SER (Society for Ecological Restoration) (2019). *What is ecological restoration?* Washington, DC: SER.

Shiferaw, B., & Holden, S. (1998). Resource degradation and adoption of land conservation technologies in the Ethiopian high lands: A case study in Andit Tid, North Shewa. *Agricultural Economics, 18*, 233–247.

Shixiong, C.S.A. (2017). A win-win strategy for ecological restoration and biodiversity conservation in Southern China. *Environmental Research Letters, 12*(4), 044004.

Sivapalan, M., Konar, M., Srinivasan, V., Chhatre, A., Wutich, A., Scott, C.A., Wescoat, J.L., & Rodríguez-Iturbe, I. (2014). Socio-hydrology: Use-inspired water sustainability science for the Anthropocene. *Earths Future, 2*(4), 225–230.

Sola, P., Oduol, J., Hagazi, N., Carsan, S., Kelly, R.Muriuki, J., Hadgu, K., & Maimbo Malesu, M. (2020). *Landscape restoration is more than land restoration: Dryland development in Ethiopia and Kenya.* Article was submitted for inclusion in the forthcoming edition of ETFRN News 60 – Restoring African drylands.

Steger, C., Hirsh, S., Evers, C., Branoff, B., Petrova, M., Nielsen-Pincus, M., & van Riper, C.J. (2018). Ecosystem Services as boundary objects for transdisciplinary collaboration. *Ecological Economics, 143*, 153–160.

Sun, G., Caldwell, P.V., & McNulty, S.G. (2015). Modelling the potential role of forest thinning in maintaining water supplies under a changing climate across the conterminous United States. *Hydrological Processes, 29*(24), 5016–5030.

Tallis, H., & Polasky, S. (2011). Assessing multiple ecosystem services: An integrated tool for the real world. In P. Kareiva, H. Tallis, T.H. Ricketts, G.C. Daily, & S. Polasky (Eds.), *Natural capital: Theory and practice of mapping ecosystem services.* New York: Oxford University Press.

Taye, G., Poesen, J,. Wesemael, B.V., Vanmaercke, M., Teka, D., Jozef Deckers, J., Goosse, T., Maetens, W., Nyssen, J., Hallet, V., & Haregeweyn, N. (2013). Effects of land use, slope gradient, and soil and water conservation structures on runoff and soil loss in semi-arid Northern Ethiopia. *Physical Geography, 34*(3), 236–259.

Teklu, B. M., Hailu, A., Wiegant, D. A., Scholten, B. S., & Van den Brink, P. J. (2018). Impacts of nutrients and pesticides from small- and large-scale agriculture on the water quality of Lake Ziway, Ethiopia. *Environmental Science and Pollution Research, 25*(14), 13207–13216.

Tenge, A.J., Graaff, J. De., & Hella, J.P. (2005). Financial efficiency of major soil and water conservation measures in West Usambara highlands, Tanzania. *Applied Geography, 25*, 348–366.

Teshome, A., Rolker, D., & de Graaff, J. (2013). Financial viability of soil and water conservation technologies in Northwestern Ethiopian highlands. *Applied Geography, 37*, 139–149.

Thakadu, O.T. (2005). Success factors in community based natural resources management in northern Botswana: Lessons from practice. *Natural Resources Forum, 29*(3), 199–212.

Tidwell, T.L. (2016). Nexus between food, energy, water, and forest ecosystems in the USA. *Journal of Environmental Studies and Sciences, 6*, 214–224.

Tilahun, A., & Belay, F. (2019). Conservation and production impacts of soil and water conservation practices under different socio-economic and bio-physical setting: A review. *Journal of Degraded and Mining Lands Management, 6* (2), 1653–1666.

Tongway, D.J., & Hindley, N.L. (2004). *Landscape function analysis: Procedure for monitoring and assessing landscapes-with special reference to Mine-sites and Rangelands.* Canberra: CSIRO Sustainable ecosystems.

UNCCD (1994). *United Nations convention to combat desertification.* Article 1. Use of terms.

UNCCD (2016). *Land degradation neutrality target setting programme land degradation neutrality target setting: A technical guide.* Bonn: UNCCD.

UNCCD and FAO (2020). *Land degradation neutrality for water security and combatting drought.* Bonn: UNCCD and FAO.

Vancampenhout, K., Nyssen, J., Gebremichael, D., Deckers, J., Poesen, J., Haile. M., & Moeyersons, J. (2006). Stone bunds for soil conservation in the Northern Ethiopian Highlands: Impacts on soil fertility and crop yield. *Soil Tillage Research, 90*, 1–15.

Wang, L., Stefano. M., Sujith. R., Diego, R.I., & Kelly, C. (2015). Dynamic interactions of ecohydrological and biogeochemical processes in water-limited systems. *Ecosphere, 6* (8), 1–27.

Wassen, M. (2007). Ecohydrology, contribution to science and practice. In T.M. de Jong, J.N.M. Dekker, & R. Posthoorn (Eds.), *Landscape ecology in the Dutch context: Nature, town and infrastructure.* Zeist, The Netherlands: KNNV Publishing.

Watson, C.C., David S. Biedenharn, D.S., & Thorne, C.R. (1999). *Demonstration erosion control: Design manual.* Vicksburg: U.S. Army Engineer Research and Development Center Vicksburg.

Wei, H., Fan, W., Ding, Z., Weng, B., Xing, K., Wang, X., Lu, N., Ulgiati, S., & Dong, X. (2017). Ecosystem services and ecological restoration in the Northern Shaanxi Loess Plateau, China, in relation to climate fluctuation and investments in natural capital. *Sustainability, 9,* 199.

West, P.C., Gerber, J.S., Engstrom, P.M., Mueller, N.D., Brauman, K.A., Carlson, K.M., Cassidy, E.S., Johnston, M., MacDonald, G.K., Ray, D.K., & Siebert, S. (2014). Leverage points for improving global food security and the environment. *Science, 345*(6194), 325–328.

Wilcox B.A., Aguirre, A.A., Padua, N.D., Siriaroonrat, B., & Echaubard, P. (2019). *Operationalizing one heath employing social-ecological systems theory: Lessons from the Greater Mekong Subregion.* Lausanne, Switzerland: Frontiers in Public Health.

Wilcox, B.P., David Le Maitre, D.L., Jobbagy, E., Wang, L., & Breshears, D.D. (2017). Ecohydrology: Processes and implications for Rangelands. In D.D. Briske (Ed.), *Rangeland Systems.* New York: Springer Series on Environmental Management.

WMO (World Meteorological Organization) (2005). *Climate and land degradation.* Geneva: WMO.

Wood, P.J., Hannah, D.M., & Sadler, J.P. (2007). *Hydroecology and ecohydrology: Past, present and future.* New York: Wiley.

Wood, P.A. (1990). Natural resource management and rural development in Ethiopia. In S. Pause Wang, F. Cheru, S. Brune, E. Chole (Eds.), *Ethiopia: Rural development options.* London and New Jersey: Zed Books Ltd.

Wordofa, M.G., Okoyo, E.N., & Erkalo, E. (2020). Factors influencing adoption of improved structural soil and water conservation measures in Eastern Ethiopia. *Environmental Systems Research, 9,* 13.

Worku, T., Khare, D., & Tripathi, S.K. (2018). Spatiotemporal trend analysis of rainfall and temperature, and its implication on crop production. *Journal of Water and Climate Change, 21,* p.jwc2018064.

Yirdaw, E., Tigabu, M., & Monge, A. (2017). Rehabilitation of degraded dryland ecosystems - review. *Silva Fennica, 51*(1B), 1673.

Zalewski, M. (2021). Ecosystem biotechnologies for the enhancement of ecohydrological potential of the catchments - water, biodiversity, ecosystem services, resilience, culture and education. *IOP Conference Series Earth and Environmental Science, 789,* 012031

Zalewski M. (2000). Ecohydrology-the scientific background to use ecosystem properties as management tools toward sustainability of water resources. *Ecological Engineering,* 16, 1–8

Zalewski, M. (2002). Ecohydrology-the use of ecological and hydrological processes for sustainable management of water resources. *Hydrological Sciences Journal, 47,* 823–832.

Zalewski, M., Santiago-Fandino, V., & Neate, J. (2003). Energy, water, plant interactions: Green feedback as a mechanism for environmental management and control through the application of phytotechnology and ecohydrology. *Hydrological Processes, 17*(14), 2753–2767.

Zalewski, M. (2006). Flood pulses and river ecosystem robustness. Frontiers in flood research. In I. Tchiguirinskaia, K. Ni Ni Thein, & P. Hubert (Eds.), *Kovacs Colloquium.* Paris: UNESCO-IAHS Publication 305.

Zalewski, M. (2010). Ecohydrology for compensation of global change. *Brazilian Journal of Biology, 70*(3), 689–695.

Zalewski, M. (2011). Ecohydrology for implementation of the UE water framework directive. *Proceedings of the Institution of Civil Engineering Water Management, 164*, 375–385.

Zalewski, M. (2013). Ecohydrology: Process-oriented thinking towards sustainable river basins. *Ecohydrology & Hydrobiology, 13*, 97–103.

Zalewski, M. (2014). Ecohydrology for engineering harmony in the changing world. In S. Eslamian (Ed.), *Handbook of engineering hydrology: Fundamentals and applications* (1st ed.). Boca Raton, FL: CRC Press. 82–94

Zalewski, M. (2015). Ecohydrology and hydrologic engineering: Regulation of hydrology-biota interactions for sustainability. *Journal of Hydrologic Engineering, 20* (1), A4014012

Zalewski, M. (2021). Ecosystem biotechnologies for the enhancement of ecohydrological potential of the catchments - Water, Biodiversity, Ecosystem Services, Resilience, Culture and Education. IOP Conf. Ser.: *Earth Environ. Sci., 789* (012031).

Zalewski, M., & Naiman, R.J. (1985). The regulation of riverine fish communities by a continuum of abiotic-biotic factors. In J.S. Alabaster (Ed.), *Habitat modification and freshwater fisheries.* London: Butterworths Scientific.

Zalewski, M. (2011). Towards engineering harmony between water, ecosystem and society: Editorial. *Ecohydrology and Hydrobiology, 11*, 137–140.

Zalewski, M., Janauer, G.A., & Jolankai, G. (1997). *Ecohydrology: A new paradigm for the sustainable use of aquatic resources* (Technical documents in hydrology No.7). Paris: UNESCO, IHP.

Zalewski, M., Santiago-Fandino, V., & Neate, J. (2003). Energy, water, plant interactions: Green feedback as a mechanism for environmental management and control through the application of phytotechnology and ecohydrology. *Hydrological Processes, 17*(14), 2753–2767.

Zalewski, M., Wagner-Lotkowska, I., & Robarts, R.D. (2004). *Integrated watershed management-ecohydrology and phytotechnology-manual.* Paris: UNESCO.

Zheng, L.D. (2004). *History of watershed resource management.* Beijing: China Water Resources Management Publishing House.

2 Guiding principles of ecohydrology-based landscape restoration (EcoLaR) approach

Mulugeta Dadi Belete

2.1 Introduction

As a fairly new approach, ecohydrology-based landscape restoration (EcoLaR) needs to have its guiding principles that denote its important values. This chapter explains those principles emanated from the basic principles of ecohydrology and amalgamated with other related approaches such as: Ecosystem Restoration (FAO et al., 2021), Forest Landscape Restoration (Newton and Tejedor, 2011), and Landscape Restoration (Sayer et al., 2013). Thus, the upcoming sub-sections describe the resulting principles.

2.2 Setting the guiding principles for ecohydrology-based landscape restoration approach

As shown in Figure 2.1, this chapter synthesizes and proposes the underlying guiding principles for the anticipated EcoLaR approach.

The first two principles guide the planning phase of landscape restoration by signifying the consideration of both hydrological and ecological management units (*principle 1*) and building local stewardship–human dimension (*principle 2*). *Principle 3* guides target-setting by indicating the multidimensional goals of the restoration efforts toward sustainability through parallel improvement of Water, Biodiversity, and Services from ecosystems for society, Resilience to climate and various anthropogenic impacts and Culture and education –WBSRCE – as proposed by Zalewski (2021). The remaining principles guide design of the management strategies in which the *fourth principle* operationalizes the first and second principles of ecohydrology by proposing regulation of the hydrology followed by place-based and use-inspired biota establishment. The *fifth principle* signifies the importance of adopting indigenous ways of doing

DOI: 10.4324/9781003309130-2

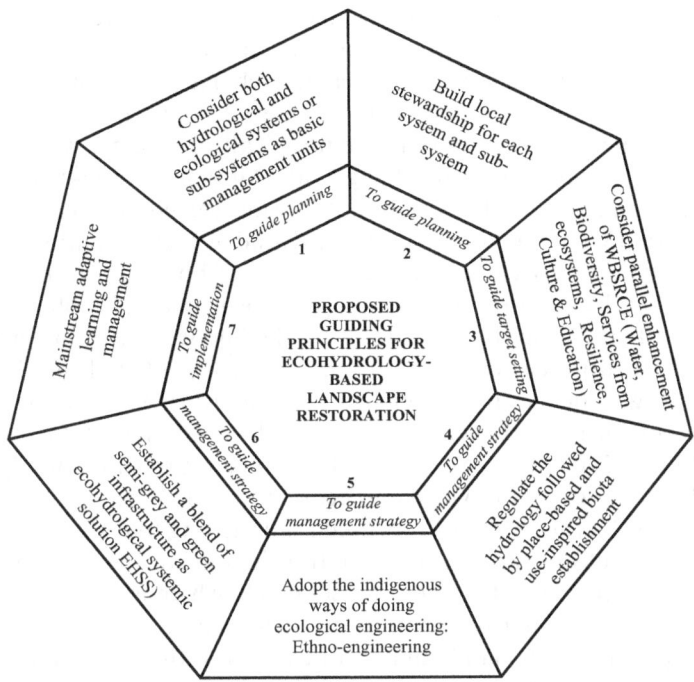

Figure 2.1 The proposed guiding principles for ecohydrology-based land-scape management.

ecological engineering: Ethno-engineering while the *sixth principle* recommends the establishment of a blended system of semi-gray and green infrastructure as ecohydrological systemic solution (EHSS). In addition, the *seventh principle* stresses the necessity of managing land-scape adaptively for emerging issues through the course of action. The upcoming sub-sections explain these principles in detail.

2.2.1 Principle 1: consider both hydrological and ecological systems and sub-systems as basic management units

This principle enables practitioners, researchers, policymakers as well as the community to perceive the landscape as both hydrologi-cal and ecological entities. Landscape is an ecological unit (Karadağ, 2013) which is composed of two or more ecosystems in close proxim-ity (Sanderson and Harris, 2000) and characterized by heterogene-ous land area with a cluster of interacting ecosystems (Forman and

Godron, 1981). Landscape approaches are primarily rooted in conservation and the science of landscape ecology (Lindenmayer et al., 2008; Sayer, 2009). It is at a higher order than the ecosystem level (Farina et al., 2005) and interwoven with climate change and ecology, development, economics, politics, and culture (Bastian et al., 2006). On the other hand, watershed is a topographically delineated area that is drained by a stream system (Wang et al., 2016) and a well-suited unit for the management of not only water resources but also ecosystems in general (Montgomery et al., 1995). Karadağ (2013) mentioned that watershed actually clarifies the complexity of boundary in the landscape. The framework of developing the principles of ecohydrology is logically the water basin scale where the importance of integration of ecological processes is most clear (Zalewski, 2002). There is also increasing evidence that watershed is an appropriate scale for planning ecological restoration (Khatami and Berndtsson, 2013; Palmer and Filoso, 2009). It is also a template for integration of knowledge on the dynamics of hydrological and ecological processes and for determining the hierarchy of derivers used for designing best strategies and systemic solutions (Zalewski, 2014). In this context, the fundamental ecological unit refers to 'an ecological patch' which is a relatively homogeneous area that differs from its surroundings (Forman, 1995). These patches in the landscape can be found in three environmental states: potential, active, and degraded (Wright et al., 2004) that call for their own restoration strategies.

2.2.2 Principle 2: build place-based landscape stewardship

Landscape approaches usually involve some form of multi-stakeholder process (Minang et al., 2015). Without the active involvement of local people and other stakeholders, restoration is unlikely to be successful (Höhl et al., 2020; Holl and Brancalion, 2020). The concept of stewardship is recently gaining popularity in the literature on resilience and sustainable social–ecological systems (SES) (Folke et al., 2016). Hence, success of any landscape approach intervention will usually depend on the actors/stakeholders within the landscape, their interests and level of engagement with the approach itself (van Noordwijk et al., 2015). Involving, supporting, and assisting communities to manage their own ecosystems are well-recognized agenda (IRC, 1993).

In the context of EcoLaR approach, landscape stewardship includes all collaborative efforts toward landscape sustainability. At the heart of this principle, there is a sense of 'deep care' for the landscape

(Bieling and Plieninger, 2017) that they are proximal to, connected to and, in some contexts, that they depend on it for subsistence needs and livelihoods (Bennett et al., 2018) and promote formation of local cultures and provide ecosystem services both for the benefit of individual and societal wellbeing (Bieling et al., 2014). It is one way through which people get involved in promoting sustainability and in responding to external derivers of change, using their own expertise and knowledge (Bennett et al., 2018). In order to maximize the benefits of the stewardship, the 'learning-practice alliance' (LPA) model of multi-stakeholders' platform is recommended in the EcoLaR approach. This LPA model is also expected to contribute to the 'culture' and 'education' components of the third principle described in the next sub-section.

2.2.3 Principle 3: consider parallel enhancement of WBSRCE (Water, Biodiversity, Services from ecosystems, Resilience, Culture, and Education)

The practice of ecological restoration has varied motivations (Wiens and Hobbs, 2015) that differ in their objectives and their methods of achieving those goals such as: restoration against fire suppression (Baker, 1994); restoration of disturbed arid landscapes (Allen, 1989), and semiarid landscapes (Ludwig and Tongway, 1996); restoration to compensate deforestation (Haggar et al., 1997); reclaiming coal surface mines (Holl and Cairns, 1994); native grassland restoration (Knapp and Rice, 1996); river restoration (Koebel, 1995; Nolan and Guthrie, 1998); retention of nutrients in river basins (Kronvang et al., 1999); restoration of an inbred adder population (Madsen et al., 1999); compensating for wetland losses (NRC, 2001); restoration of disturbed ecosystems (Palik et al., 2000); restoring habitats (Schultz and Crone, 1998); restoration of abandoned rangeland (Scowcroft and Jeffrey, 1999); conservation by restoration (Tockner et al., 1998); carbon, nitrogen, and phosphorus cycling in wetlands (Van der Peijl and Verhoeven, 2000); restoration and sustainable development of mining areas (Lei et al., 2016); and many others. Most of these interventions take too long, cost too much, and produce too few benefits to justify public or private expenditures (Verdone and Seidl, 2017). The concept of providing multiple ecosystems from landscape restoration activities is gaining momentum these days. Yet, we still lack much of the knowledge needed to operationalize and implement restoration successfully while also addressing the needs and aspirations of landholders (Chazdon et al., 2015).

In the face of crisis generated by degradation and overexploitation of Biosphere, we need to target and find innovative solutions for parallel/ simultaneous enhancement/improvement of key parameters which determine catchment sustainability, WBSRCE (Figure 2.2) (Zalewski, 2021), parameters that need to be simultaneously improved by means of regulation of EH processes (Zalewski, 2014). These dimensions are societal needs that need to be harmonized with the enhanced ecosystems potential (Zalewski, 2002; Zalewski and Robarts, 2003), and it is said to be impossible to achieve sustainable growth without addressing these six dimensions (Jarosiewicz et al., 2021). Such a multidimensional goal for landscape management is the fundamental condition for the transition from exploitative to sustainable resource use, incorporating innovative tools as 'Nature-Based Solutions' and a systemic approach compatible with the 'Circular and Bio-Economy' (Zalewski, 2014).

2.2.4 Principle 4: regulate the hydrology followed by place-based and use-inspired biota establishment

One of the most important (though poorly recognized by decision makers and practitioners) consequences of our present policies is acceleration of the hydrological cycle. 'Regulation' is a proactive management of ecosystems and intrinsically related to ABRC which gave birth to the science of ecohydrology (EH) and became a key word for process-oriented thinking of ecohydrology. It has been appearing as a step toward the new integrative environmental science in which the ultimate goal is sustainability (Zalewski, 2014). The concept of EH was formulated as a response to the need for developing a methodology on how to regulate hydrological cycle at various spatial scales toward sustainability (Zalewski et al., 1997; Zalewski, 2002).

The 'EcoLaR' approach comprises different ecohydrological solutions that basically provide mechanisms controlling water fluxes toward reducing the stochastic variation of water dynamics and their consequences, e.g., erosion and accelerated nutrient transfer across landscape gradients (Zalewski, 2002). By doing this, we are regulating (stabilizing) the dominant abiotic processes (hydrology) that in turn facilitate biotic interactions start to manifest themselves (Zalewski and Naiman, 1985). Jørgensen (2016) also noted that the control of hydrological retention time is a prerequisite for the application of EH in environmental management. Through such hydrologic regulations, the paths taken by water is going to be managed, which directly or indirectly determine many of the characteristics of a landscape, the occurrence and size of floods, the uses to which land may be put, and the strategies required for wise land management (Tarboton, 2003).

Since water is a major deriver of biogeochemical evolution, and hence of biodiversity and biological productivity, regulation of ecohydrological processes becomes the first and fundamental step toward achieving sustainability in the catchment, particularly in human-modified and degraded systems (Zalewski, 2014). In ecohydrologic sense, regulated hydrology is a prerequisite for stable ecosystem services such as flood control, erosion control, sediment control, etc. Hence, the practice of EcoLaR approach strives to keep the runoff as sheet or overland flow to avoid significant soil erosion. Due to this, the water in the form of runoff is to be managed as close to its sources as much as possible.

According to the *first principle of ecohydrology* (H1), the hydrological framework serves as a template for functional integration of hydrological and biological processes (Zalewski et al., 1997). If the hydrologic processes are not stable, we do not expect biotic interactions to start manifesting themselves (Zalewski, 2002). Furthermore, establishing vegetation on severely degraded landscapes is difficult because of the strong erosive forces, especially in dry climates and on poor soils (Stokes et al., 2014). In such environments, plant growth is often controlled by stochastic pulses of water that directly affect plants' ability to adapt and survive (Schwinning and Sala, 2004). We also know that plants need water to survive, and thus, the distribution, composition, and structure of plant communities are directly influenced by spatiotemporal patterns in water availability (Asbjornsen et al., 2011). In other words, the establishment and growth of vegetation depends on the availability of soil water resources (D'Odorico and Porporato, 2006).

2.2.5 Principle 5: adopt the indigenous ways of doing ecological engineering: Ethno-engineering

These days, engineering as a practice is contributing to social justice (Leydens et al., 2012; Riley, 2008) and responsible development (Roco, 2005). For the sake of sustainability, these practices need to consider ethno-engineering solutions that refer to the indigenous or locally contextualized ways of doing engineering. Some of the characteristics of ethno-engineering solutions include dominant use of local resources (as opposed to imported materials and technologies) and high levels of sustainability (Hess and Strobel, 2013). This approach is also considered as holistic and adaptive by nature and acknowledges the ever-changing environmental conditions (Wang et al., 2016).

2.2.6 Principle 6: establish a blended system of green and semi-gray infrastructure as ecohydrological systemic solution (EHSS)

Creating a scientific basis for a socially acceptable, cost-effective, and systemic approach to the sustainable management of freshwater resources is one of the key components of ecohydrologic programs (IHP, 2013). Green-(semi-) gray infrastructure is a strategically planned network of natural and semi-natural features that are designed or managed to deliver a wide range of ecosystem services (Estreguil et al., 2019) and, in our case, it refers to the renovated ecological engineering measures that are able to capture, retain, remediate, and slowly release water that might otherwise have passed rapidly through an area.

The concept of green- (semi-) gray infrastructure has been a popular framework for smart development and conservation planning (Chang et al., 2012) that entails 'multi-functionality' and 'connectivity', while the ecohydrology concept seeks to manage water resources through managing dual regulation of hydrology and biota (Mitsch, 1993; Zalewski et al, 1990). The multi-functionality principle is also shared by landscape restoration that acknowledges reconciliation of stakeholders' multiple needs, preferences, and aspirations of the diverse range of values, goods, and services derived from the restoration efforts (Sayer et al., 2013). On the other hand, ecohydrology strives to utilize such processes for enhancing environmental sustainability (Zalewski, 2010) in which the biocenotic processes are shaped by hydrology and, vice versa, and the biocenotic structure and interactions shape hydrological processes (Zalewski, 2000; 2006). Such management is perceived to improve the coexistence of man and nature by enhancing the absorbing capacity of ecosystems (Zalewski, 2002).

In practice, the concept of green- (semi-) gray infrastructure ranges from small-scale technologies to regional planning strategies targeting conservation or restoration of natural landscapes and watersheds (US-EPA, 2017) to deliver a wide range of ecosystem services in both rural and urban settings (EC, 2013). This concept uses soil and vegetation to utilize, enhance, and/or mimic the natural hydrologic cycle processes of infiltration, evapotranspiration, and reuse (US-EPA, 2008) with multifunctional character and lower environmental cost (Caparrós et al., 2020).

2.2.7 Principle 7: adaptive learning, management, replication, and scaling-up

Adaptive management (AM) in landscape restoration efforts can improve the ability to cope with the inherent uncertainties of managing complex dynamic systems by learning from the outcomes of management implementation and adjusting future approaches accordingly (Allan et al., 2008; Porzecanski et al., 2012). This principle refers to the need for encompassing a series of actions characterized by feedback loops, with deliberate intent to achieve goals through the modification and refinement of hypotheses, objectives, outputs or outcomes, and of management actions (Kingsford et al., 2017). Adaptive management is proposed to the EcoLaR approach in response to the fact that ecohydrology is an emerging science calling for a learning-by-doing paradigm. Through the course of action, practitioners may adjust, improve existing routines' (single-loop learning), 'reframe and change practice' (double-loop learning), and 'review norms and values, and transform governance' (triple-loop learning) (Fabricius and Cundill, 2014; Pahl-Wostl, 2009).

2.3 The EcoLaR principles in view of the previous ones

As presented in Figures 2.2–2.4, the principles proposed for the EcoLaR approach address a fairly wider scope than the available

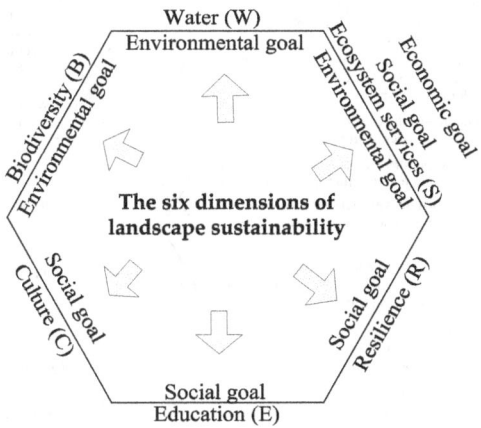

Figure 2.2 The proposed multidimensional goals (key elements of the ecohydrology paradigm).

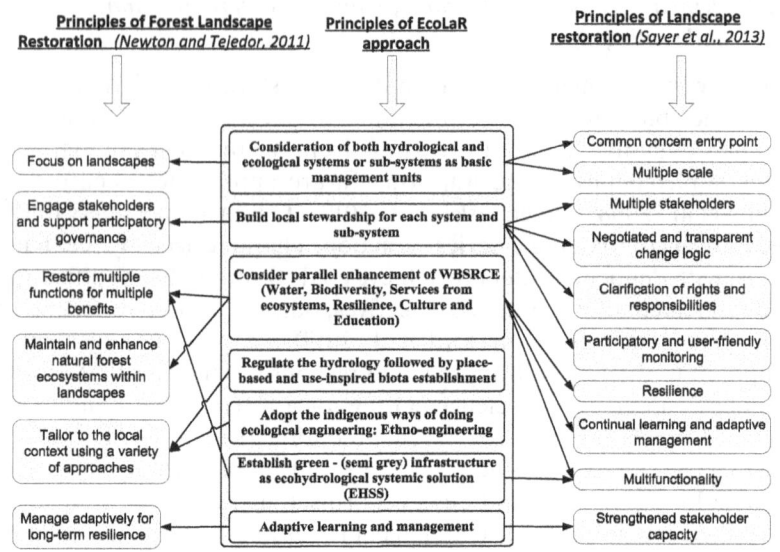

Figure 2.3 Comparison of EcoLaR principles with (forest) landscape restoration principles.

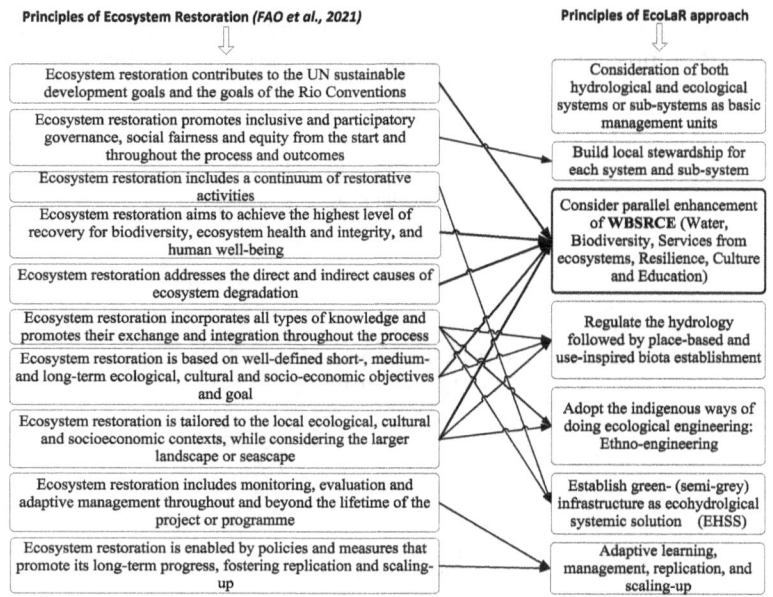

Figure 2.4 Comparison of EcoLaR principles with ecosystem restoration principles.

principles in the sector such as the landscape restoration principles of Sayer et al. (2013); forest landscape restoration principles of ITTO (2020), and the more recent ecosystem restoration principles of FAO, IUCN CEM & SER (2021). From the figures, it is easy to recognize the particularity of the third principle to entail the critical sustainability dimensions (Figure 2.2).

References

Allan, C., Curtis A., Stankey G., & Shindler, B. (2008). Adaptive management and watersheds: A social science perspective. *Journal of the American Water Resources Association*, *44*(1), 166–174.

Allen, E.B. (1989). The restoration of disturbed arid landscapes with special reference to mycorrhizal fungi. *Journal of Arid Environments, 17*, 279–286.

Asbjornsen, H., Goldsmith, G.R., Alvarado-Barrientos, M.S., Rebel, K., Van Osch, Rietkerk, M., Chen, J., Gotsch, S., Tobón, C., Geissert, D.R., Gómez-Tagle, A., Vache, K., & Dawson, T.E. (2011). Ecohydrological advances and applications in plant-water relations research: A review. *Journal of Plant Ecology, 4*(1–2), 3–22.

Baker, W.L. (1994). Restoration of landscape structure altered by fire suppression. *Conservation Biology, 8*, 763–769.

Bastian, O., Krönert, R., & Lipsky Z. (2006). Landscape diagnosis on different space and time scales – A challenge for landscape planning. *Landscape Ecology, 21*, 359–374.

Bennett, A., Patil, P., Kleisner, K., Rader, D., Virdin, J., & Basurto, X. (2018). *Contribution of fisheries to food and nutrition security: Current knowledge, policy, and research* (NI Report 18-02). Durham, NC: Duke University.

Bieling, C., & Plieninger, T. (2017). *The science and practice of landscape stewardship*. Cambridge: Cambridge University Press.

Bieling, C., Plieninger, T., Pirker, H., & Vogl, C.R. (2014). Linkages between landscapes and human well-being: An empirical exploration with short interviews. *Ecological Economics, 105*, 1–30.

Caparrós, J.L., Juan, M.G., Nuria, R.L., & Jaime, D.P. (2020). Green Infrastructure and Water: An Analysis of Global Research. *Water, 12*, 1760.

Chang, Q., Li, X., Huang, X., & Wu, J. (2012). A GIS-based green infrastructure planning for sustainable urban land use and spatial development. *Procedia Environmental Sciences, 12*, 491–498.

Chazdon, R.l., Brancalion, P.H.S., Lamb, D., Laestadius, L., Calmon, M., & Kumar, C. (2015). A policy-driven knowledge agenda for global forest and landscape restoration. *Conservation Letter, 10*, 125–132.

D'Odorico, P., & Porporato, A. (2006). *Dryland ecohydrology*. The Netherlands: Springer.

EC (European Commission) (2013). *Communication from the Commission to the European Parliament, the Council, the European Economic and Social Committee and the Committee of the Regions Green Infrastructure*

(GI)-Enhancing Europe's Natural Capital. Retrieved from http://eurlex.europa.eu/LexUriServ/LexUriServ.do?uri=COM:2013:0249:FIN:EN:PDF.

Estreguil, C., Dige, G., Kleeschulte, S., Carrao, H., Raynal, J., & Teller, A. (2019). *Strategic green infrastructure and ecosystem restoration: Geospatial methods, data and tools* (EUR 29449 EN). Luxembourg: Publications Office of the European Union.

Fabricius, C., & Cundill, G. (2014). Learning in adaptive management: Insights from published practice. *Ecology and Society, 19*(1), 29.

FAO, IUCN CEM & SER (2021). *Principles for ecosystem restoration to guide the United Nations Decade 2021–2030.* Rome: FAO.

Farina, A., Bogaert, J., & Schipani, I. (2005). Cognitive landscape and information: New perspectives to investigate the ecological complexity. *Bio Systems, 79,* 235–240.

Folke, C., Biggs, R., Norström, A.V., Reyers, B., & Rockström, J. (2016). Social-ecological resilience and biosphere-based sustainability science. *Ecology and Society, 21*(3), 41.

Forman, R.T.T. (1995). *Land mosaics: The ecology of landscapes and regions.* Cambridge, UK: Cambridge University Press.

Forman, R.T.T., & Godron, M. (1981). Patches and Structural Components for a Landscape Ecology. *BioScience, 31*(10), 733–740.

Haggar, J., Wightman, K., & Fisher, R. (1997). The potential of plantations to foster woody regeneration within a deforested landscape in lowland Costa Rica. *Forest Ecology and Management, 99,* 55–64.

Hess, J., & Strobel, J. (2013). Sustainability and the engineering worldview. *Proceedings - Frontiers in Education Conference,* 644–648. Oklahoma City, OK, USA.

Höhl, M., Ahimbisibwe, V., Stanturf, J.A., Elsasser, P., Kleine, M., & Bolte, A. (2020). Forest landscape restoration-what generates failure and success? *Forests, 11,* 938.

Holl, K.D. & Cairns, J. Jr. (1994). Vegetational community development on reclaimed coal surface mines in Virginia. *Bulletin of the Torrey Botanical Club, 121,* 327–337.

Holl, K.D., & Brancalion, P.H.S. (2020). Tree planting is not a simple solution. *Science, 368,* 580–581.

IHP (2013). *Ecohydrology for Sustainability.* Paris: UNESCO.

IRC (1993). *Community management today – The role of communities in the management of improved water supply systems.* The Hague:International Water and Sanitation Centre. ITTO (International Tropical Timber Organization) (2020). *Guidelines for forest landscape restoration in the tropics.* ITTO Policy Development Series No. 24. Yokohama, Japan.

Jarosiewicz, P., Jurczak, T., & Zalewski, M. (2021). *Ecohydrology for sustainable urban water management. Pre-conference for the second international conference on water, megacities and global change.* UNESCO. Paris.

Jørgensen, S.E. (2016). Ecohydrology as an important concept and tool in environmental management. *Ecohydrology & Hydrobiology, 16*(1), 4–6.

Karadağ, A.A. (2013). Use of watersheds boundaries in the landscape planning. In M. Özyavuz (Ed.), *Advances in landscape architecture*. London: IntechOpen.

Khatami, S., & Berndtsson, R. (2013). *Urmia lake watershed restoration in Iran: Short- and long-term perspectives*. 6th International Perspective on Water Resources & the Environment conference (IPWE 2013), Environmental & Water Resources Institute (EWRI)/ American Society of Civil Engineers (ASCE), Izmir: Turkey.

Kingsford, R.T., Roux, D.J., McLoughlin, C.A., Conallin, J., & Norris, V. (2017). Strategic adaptive management (SAM) of intermittent rivers and ephemeral streams (IRES) - Juggling temporal and spatial scales, cultures and governance. In T. Datry, N. Bonada, & A. Boulton (Eds.), *Intermittent rivers and ephemeral streams: Ecology and management*. Burlington, VT: Academic Press.

Knapp, E.E., & Rice, K.J. (1996). Genetic structure and gene flow in Elymus glaucus (blue wildrye): Implications for native grassland restoration. *Restoration Ecology, 4*, 1–10.

Koebel, J.W. (1995). An historical perspective on the Kissimmee River restoration project. *Restoration Ecology, 3*, 149–159.

Kronvang, B., Hoffmann, C.C., Svendsen, L.M., Windolf, J., Jensen, J.P., & Dorge, J. (1999). Retention of nutrients in river basins. *Aquatic Ecology, 33*, 29–40.

Lei, K., Pan, H., & Lin, C. (2016). A landscape approach towards ecological restoration and sustainable development of mining areas. *Ecological Engineering, 90*, 320–325.

Leydens, J.A., Lucena, J.C., & Schneider, J. (2012). Are engineering and social justice (in) commensurable? A theoretical exploration of macrosociological frameworks. *International Journal of Engineering Social Justice and Peace, 1*, 63–82.

Lindenmayer, D., Hobbs, R.J., Montague-Drake, R., Alexandra, J., Bennett, A., Burgman, M., Cale, P., Calhoun, A., Cramer, V., Cullen, P., Driscoll, D., Fahrig, L., Fischer, J., Franklin, J., Haila, Y., Hunter, M., Gibbons, P., Lake, S., Luck, G., MacGregor, C., McIntyre, S., Nally, R.M., Manning, A., Miller, J., Mooney, H., Noss, R., Possingham, H., Saunders, D., Schmiegelow, F., & Scott, M. (2008). A checklist for ecological management of landscapes for conservation. *Ecology Letter, 11*, 8–91.

Ludwig, J.A., & Tongway, D.J. (1996). Rehabilitation of semiarid landscapes in Australia. II. Restoring vegetation patches. *Restoration Ecology, 4*, 398–406.

Madsen, T., Shine, R., Olsson, M., & Wittzell, H. (1999). Restoration of an inbred adder population. *Nature, 402*, 34–35.

Minang, P.A., Duguma, L.A., Van Noordwijk, M., Prabhu, R., & Freeman, O.E. (2015). Enhancing multi-functionality through system improvement and landscape democracy processes: A synthesis. In P.A. Minang, M. Van Noordwijk, O.E. Freeman, C. Mbow, J. De Leeuw, D. Catacutan (Eds.),

Climate-smart landscapes: Multifunctionality in practice. Nairobi: World Agroforestry Centre (ICRAF).

Mitsch, W. (1993). Ecological Engineering-a co-operative role with planetary life-support system. *Environmental Science and Technology, 27*, 438–445.

Montgomery, D.R., Grant, G.E., & Sullivan, K. (1995). Watershed analysis as a framework for implementing ecosystem management. *Water Resources Bulletin, 31*(3), 369–386.

Newton, A.C., & Tejedor, N. (Eds.). (2011). *Principles and practice of forest landscape restoration: Case studies from the drylands of Latin America.* Gland: IUCN.

Nolan, P.A., & Guthrie, N. (1998). River rehabilitation in an urban environment: Examples from the Mersey Basin, North West England. *Aquatic Conservation, 8*, 685–700.

NRC (National Research Council) (2001). *Compensating for wetland losses under the clean water act.* Washington, DC: National Academy Press.

Pahl-Wostl, C. (2009). A conceptual framework for analysing adative capacity and mul-level learning processes in resource governance regimes. *Global Environmental Change, 19*(3), 354–365.

Palik, B.J., Goebel, P.C., Kirkman, L.K., & West, L. (2000). Using landscape hierarchies to guide restoration of disturbed ecosystems. *Ecological Applications, 10*, 189–202.

Palmer, M.A., & Filoso, S. (2009). Restoration of ecosystem services for environmental markets. *Science, 325*, 575.

Porzecanski, I., Saunders, L.V., & Brown, M.T. (2012). Adaptive management fitness of watersheds. *Ecology and Society, 17*(3), 29–43.

Riley, D. (2008). *Engineering and social justice.* San Rafael, CA: Morgan & Claypool.

Roco, M.C. (2005). Environmentally responsible development of nanotechnology. *Environmental Science and Technology, 39*(5), 106A–112A.

Sanderson, J., & Harris, L.D. (Eds.). (2000). *Landscape ecology: A top-down approach.* Boca Raton, FL: Lewis Publishers.

Sayer, J. (2009). Reconciling conservation and development: Are landscapes the answer?. *Biotropica, 41*, 649–652.

Sayer, J., Terry Sunderland, T., Ghazoul, J., Jean-Laurent Pfund, J.L., Douglas Sheil, D., Meijaard, E., Venter, M., Boedhihartono, A.K., Day, M., Claude Garcia, C., Cora van Oosten, C., & Buck, L.E. (2013). Ten principles for a landscape approach to reconciling agriculture, conservation, and other competing land uses. *Proceedings of the National Academy of Sciences, 110*, 8349–8356.

Schultz, C.B., & Crone, E.E. (1998). Burning prairie to restore butterfly habitat: A modeling approach to management tradeoffs for the Fender's blue. *Restoration Ecology 6*, 244–252.

Schwinning, S., & Sala, O.E. (2004). Hierarchy of responses to resource pulses in and semi-arid ecosystems. *Oecologia, 141*, 211–220.

Scowcroft, P.G., & Jeffrey, J. (1999). Potential significance of frost, topographic relief, and Acacia koa stands to restoration of mesic Hawaiian forests on abandoned rangeland. *Forest Ecology and Management, 114,* 447–458.

Stokes, A., Douglas, G.B., Fourcaud, T., Giadrossich, F., Gillies, C., Hubble, T., Kim, J.H., Loades, K.W., Mao, Z., McIvor, I.R., Mickovski, S.B., Mitchell, S., Osman, N., Phillips, C., Poesen, J., Polster, D., Preti, F., Raymond, P., Rey, F., Schwarz, M., & Walker, L.R. (2014). Ecological mitigation of hillslope instability: Ten key issues facing researchers and practitioners. *Plant Soil, 377,* 1–23.

Tarboton, D.G. (2003). *Rainfall-runoff processes: A workbook to accompany the rainfall-runoff processesweb module.* Utah: Utah State University.

Tockner, K., Schiemer, F., & Ward, J.V. (1998). Conservation by restoration: The management concept for a river-floodplain system on the Danube River in Austria. *Aquatic Conservation, 8,* 71–86.

US-EPA (2008). *Managing wet weather with green infrastructure. Action strategy.* Washington, DC: United States Environmental Protection Agency.

US-EPA (2017). What is Green Infrastructure? United States Environmental Protection Agency. Retrieved from: https://www.epa.gov/green-infrastructure/what-green-infrastructure

Van der Peijl, M.J., & Verhoeven, J.T.A. (2000). Carbon, nitrogen and phosphorus cycling in river marginal wetlands: A model examination of landscape geochemical flows. *Biogeochemistry, 50,* 45–71.

van Noordwijk, M., Minang, P.A., & Hairiah, K. (2015). Shifting cultivation in an era of climate change. In M. Cairns (Ed.), *Shifting cultivation and environmental change: Indigenous people, agriculture and forest conservation.* Oxfordshire, UK: Routledge.

Verdone, M., & Seidl, A. (2017). Time, space, place, and the Bonn Challenge global forest restoration target. *Restoration Ecology, 25* (6), 903–911.

Wang, G., Mang, S., Cai, H.; Liu, S., Zhang, Z., Wang, L., & Innes, J.L. (2016). Integrated watershed management: evolution, development and emerging trends. *Journal of Forestry Research, 27,* 967–994.

Wiens, J.A., & Hobbs, R.J. (2015). Integrating conservation and restoration in a changing world. *BioScience, 65,* 302–312.

Wright, J.P., Gurney, W.S.C. & Jones, C.G. (2004). Patch dynamics in a landscape modified by ecosystem engineers. *Oikos, 105,* 336–348.

Zalewski, M. (2000). Ecohydrology - The scientific background to use ecosystem properties as management tools toward sustainability of water resources. *Ecological Engineering, 16,* 1–8.

Zalewski, M. (2002). Ecohydrology, the use of ecological and hydrological processes for sustainable management of water resources. *Hydrological Sciences-Journal des Sciences hydrologiques, 47*(5), 823–832.

Zalewski, M. (2006). Ecohydrology - An interdisciplinary tool for integrated protection and management of water bodies. *Large Rivers, 158*(4), 613–622.

Zalewski, M. (2010). Ecohydrology for compensation of Global Change. *Brazilian Journal of Biology, 70* (3), 689–695.

Zalewski, M. (2014). Ecohydrology, biotechnology and engineering for cost efficiency in reaching the sustainability of biogeosphere. *Ecohydrology & Hydrobiology, 14*(1), 14–20.

Zalewski, M. (2021). Ecosystem biotechnologies for the enhancement of eco-hydrological potential of the catchments – Water, biodiversity, ecosystem services, resilience, culture and education. *IOP Conference Series Earth and Environmental Science, 789*, 012031.

Zalewski, M., & Naiman, R.J. (1985). The regulation of riverine fish communities by a continuum of abiotic-biotic factors. pp. 3–9. In J.S. Alabaster (Ed.), *Habitat modification and freshwater fisheries.* London: FAO/UN/Butterworths Scientific.

Zalewski, M., Brewinska-Zaras, B., Frankiewicz, P., & Kalinowski, S. (1990). The potential for biomanipulation using fry communities in a lowland reservoir: Concordance between water quality and optimal recruitment. *Hydrobiologia, 200/201*, 549–556.

Zalewski, M., Janauer, G. A., & Jolankai, G. (Eds). (1997). *Ecohydrology. A new paradigm for the sustainable use of aquatic resources.* Paris: UNESCO IHP.

3 Conceptual design of green-(semi-) gray infrastructure as ecohydrological systemic solution for landscape restoration

Mulugeta Dadi Belete

3.1 Introduction

Landscape restoration is not simply planting trees (Chazdon and Brancalion, 2019), but generally refers to the process of assisting the recovery of an ecosystem that has been degraded, damaged, or destroyed (SER, 2004) and plays a central role in the provision of ecosystem services and realization of the UN's Sustainable Development Goals (Yirdaw et al., 2017). It is also more than land restoration (Sola et al., 2020) and usually targets the reparation of ecosystem processes, productivity, and services without necessarily achieving a return to 'pre-disturbance' conditions (CBD, 2012; Mansourian, 2005). Sustainable development, on the other hand, is a well-established concept, but there are still disputes on how and what would be the best way to achieve it thus, new strategies and measures for achieving it are urgently needed (Zalewski, 2014). However, there is remarkable lack of appropriate methods to achieve this seriously sought-after goal called sustainable development. At this level of understanding, we all agree that the hydrotechnical solutions should not be destructive and should rely on the principle of minimum disturbance on the ecosystem. In this regard, the 'EcoLaR' approach brings together different ecohydrological solutions that basically provide mechanisms controlling water fluxes toward reducing the stochastic variation of water dynamics and their consequences such as erosion and accelerated nutrient transfer across landscape gradients (Zalewski, 2002).

Thus, the use of ecohydrologic strategy remains a relatively new and important strategy that environmental managers are increasingly interested in, but, so far, the link between theory and practice of landscape restoration is limited. This chapter strives to conceptually design a landscape scale system of green- (semi-) gray infrastructure

DOI: 10.4324/9781003309130-3

based on the theory of ecohydrology (EH). In this regard, the underlying ecohydrological functionality of the proposed system is dictated by two concerns: 'how to achieve hydrologic regulation' in line with the *first ecohydrologic principle*, on one hand, and 'how to provide ecologically favorable conditions for place-based and use-inspired biota growth' which is in line with the *second ecohydrologic principle* on the other. Controlling the hydrological retention time is noted as a prerequisite for the application of EH in environmental management in the sense that it is a prerequisite for all life (Jørgensen, 2016). Once the above design issues are satisfied, the landscape is expected to ecohydrologically restored so as to perform the 'water-biota interactions' in line with the *third principle of EH* that can lead to sustainability. Such environmental management tool will deliver multiple ecosystem services that eventually contribute to the livelihood of the community beyond their ecohydrological roles.

3.2 Conceptualizing the basic structural unit of the anticipated green- (semi-) gray infrastructure

In line with the above design principles, the basic structural unit of the proposed infrastructure fulfills 'regulation of overland flow' using ecological engineering technique (Figure 3.1). In the EcoLaR approach, the design of this unit is oriented by the ecological engineering design principles proposed by Bergen et al. (2001). Water is the most valuable natural resource and its availability and quality are essential for the proper functioning of essential ecosystems (IHP, 2013). To this end, the proposed basic structural unit regulates the hydrologic cycle by influencing the run-off down the slope.

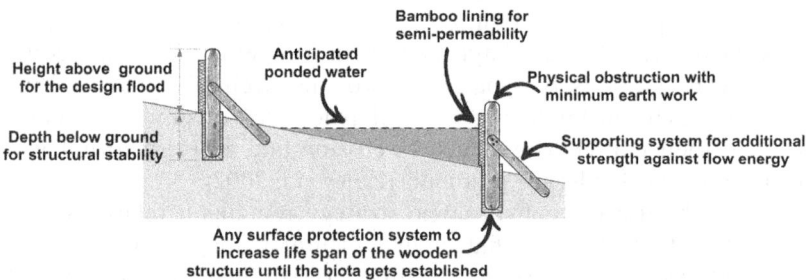

Figure 3.1 Schematics of the basic resource regulating system (not to scale).

From ecological engineering perspective, this basic structural unit of the green- (semi-) gray infrastructure comprises a sequence of semi-permeable wooden barriers that act as obstructions for slowing down and capturing overland flow that in turn increases water retention capability of the landscape. The unit regulates the key hydrological parameters such as flow velocity and concentration through which hydrological processes such as run-off and infiltration are to be regulated (*first principle of EH*). Ecologically, this unit creates abiotic template for biotic manifestation that mimics natural pools for resources accumulation. This abiotic template in turn contributes to the regulation of fundamental ecological process such as water and nutrients cycling in the ecosystem which is closely linked to water availability (Sun et al., 2017). In this regard, the sequential operation of this basic unit enables the landscape to retain, utilize, and cycle resources that characterize a healthy (functional) landscape as opposed to the dysfunctional landscape that tends to lose its vital resources (Ludwig and Tongway, 1997).

Generally, this basic unit is based on the three principles of EH (Zalewski, 2011; Zalewski et al., 1997) and potentially addresses the six dimensions of sustainable development: WBSRCE (Water, Biodiversity, Services from ecosystems, Resilience, Culture, and Education) (Zalewski, 2015). For the *W* (*water*) component, the unit provides multiple hydrological functionalities such as increment in water-capturing and retention capability of the landscape through which soil moistures enhancement, groundwater recharge, slowing down the transfer of water, and water quality enhancement. The unit also contributes to the *B* (*biodiversity*) dimension by enhancing the diversity of critical terrestrial habitats through its regulatory role on water, energy, and nutrients circulation which in turn maintains biodiversity. Through its multiple ecosystem services, the unit also addresses the *S* (*services from the ecosystem*) dimension such as regulation against flood and drought events, and provision of food items in terms of biota response following regulation of the hydrology. *The R* (*resilience*) dimension will also be potentially addressed through capability of this basic unit to control the threats of flood and drought as well as enhancement of soil moisture as an opportunity toward climate-smart landscape. The unit also directly or indirectly contributes to the *CE* (*culture and education*) through its capability to maximize indigenous knowledge as well as ethno-engineering principles that create wide opportunities to develop citizens' science for sustainable future.

3.3 Mechanism to fulfill the first ecohydrological principle as a framework: regulating the hydrology for biota growth

Water is the most important element in ecosystems (Changming et al., 2009*)*. In the terrestrial phase of EH, the basic eco-engineering unit basically requires a deep understanding of water–plant–soil interactions (Eamus et al., 2006; Guswa et al., 2020) as water is the very bloodstream of the life-support system on the planet (Ripl, 2003). It is crucial for catchment management, especially for shaping terrestrial ecosystems distribution and structures for concordant enhancement of resilience to climate change (Zalewski, 2021). This basic system mimics natural systems where the transfer of vital resources among patches, representing losses from donor ecosystems and subsidies to recipient ecosystems are regulated, for the long-term sustainability of ecosystems (Carpenter et al., 1999; Chapin et al., 2002).

This ecologically engineered unit basically regulates the 'abiotic' environment (the hydrology) that potentially alters the ecosystem structure and facilitates 'biotic' changes and feedbacks. It provides a mechanism that controls water fluxes toward reducing the stochastic variation of water dynamics and their consequences, such as erosion and accelerated nutrient transfer across landscape gradients (Zalewski, 2002). By doing this, we are regulating (stabilizing) the dominant abiotic processes (hydrology) that in turn triggers self-manifestation of the biotic feedback (Zalewski and Naiman, 1985). Manifestation of the 'biota' following the stable and predictable 'hydrology' is governed by the notion of 'hierarchy of factors' that underlies the concept of EH (Zalewski and Naiman, 1985). This two-way (dual) interaction is the basic essence of EH as expressed by one of the key hypotheses that 'The regulation of hydrological parameters in an ecosystem or catchment can be applied to control biological processes' (Zalewski, 2000).

By virtue of its flow regulation effect (Figure 3.2), the run-off produced between two consecutive barriers will gain an extended residence time so as to infiltrate into the soil and enhance soil moisture content as well as recharge the groundwater system across the landscape. This infiltration process is the critical pivot point for biota response especially in water-limited terrestrial ecosystem. In this regard, this basic structural unit as well as the entire system operates to slow down the transfer of water from the atmosphere to lakes or other water bodies which is among the major goals of EH as a problem-solving science. Once the hydrology (the driving force) is regulated, the subsequent regulation of nutrients, sediments, and organic matter will follow.

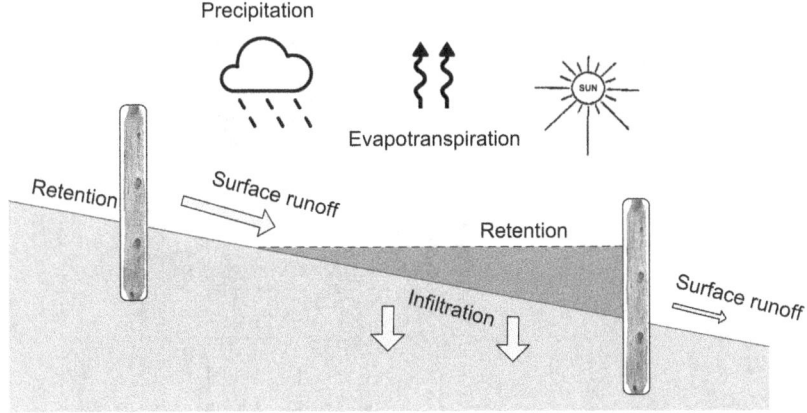

Figure 3.2 Conceptual framework to demonstrate how the anticipate green-(semi-) gray infrastructure operates toward the fulfillment of the first principle of EH.

3.4 Mechanism to fulfill the second ecohydrological principle as a target: shaping the biota to regulate the hydrology

This regulating mechanism of the proposed system (Figure 3.2) is shaped by the assumption that signifies the importance of abiotic factors (the hydrology, in this case) being the primary factor to control the biotic interactions (plant growth) (Figure 3.3). Once the hydrology becomes stable and predictable, the biotic interactions start to manifest themselves (Zalewski and Naiman, 1985). Such regulation, in turn, slows down transfer of water in the landscape. The general assumption is that water is determinant of carbon retention in terrestrial ecosystems, biomass and plant production, and ecological succession (Zalewski, 2002). Such understanding of the ecological responses in the catchment is necessary to create a comprehensive plan for catchment management in terms of protection, rehabilitation, and management (Jarosiewicz et al., 2021). Kundzewicz et al. (2002) noted the active role of plants in modifying the hydrological cycle and much of the water fluxes such as transpiration and evapotranspiration, interception, stemflow, throughfall, sediment transport, soil water depletion, water intake, plant water use, and water quality to have explicit links with vegetation. Vegetation can also exert a positive effect on infiltration rates and soil hydraulic conductivity due to organic matter

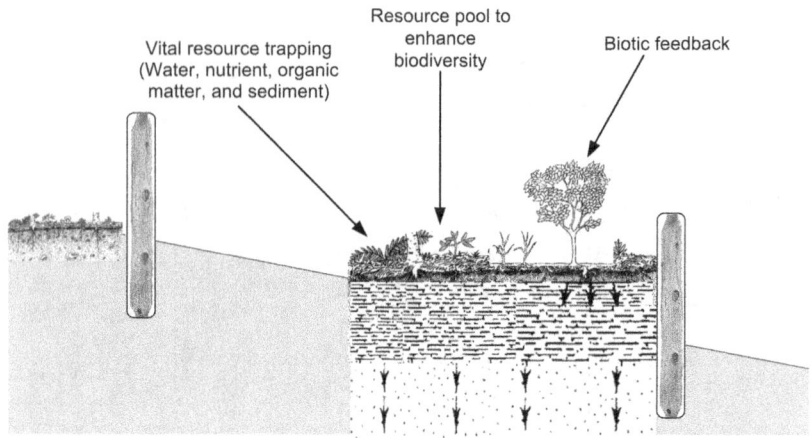

Figure 3.3 Conceptual framework to demonstrate how the anticipated system operates toward the fulfillment of the second principle of EH.

accumulation, increased root activity, and improved physical properties (Bonell et al., 2010; Germer et al., 2010). This affects overland water flow, source–sink relationships, and plant productivity (Ludwig et al., 2005; Popp et al., 2009; Reid et al., 1999).

From the perspectives of ecological principle of EH, the proposed system is intended to regulate ecosystem structure and processes toward increasing the 'carrying capacity' (water quality, restoration of biodiversity, ecosystem services for society, and resilience of river ecosystem) (Vorosmarty and Sahagian, 2000; Kedziora and Ryszkowski, 1999). The carrying capacity is attributed to the diversified plant biomass which efficiently reduces leakage of nutrients from the system. In parallel with this, the system is meant to restore and maintain critical habitats for water, energy and nutrients circulation toward sustainability (Harper et al., 2008). According to the ecological principle of EH, the feedbacks from biotic processes impact the water cycle (Newman et al., 2006). In the terrestrial EH, the water–plant–soil/ground water interactions and plant cover are the first important filtering systems that also enhance the infiltration and stabilization of water circulation within the catchment (Zalewski, 2013). In relation to societal responses, the proposed system also solves immediate societal threats by contributing to efforts of flood control, soil erosion control, and protecting fresh water bodies from siltation problem.

3.5 Fulfillment of the third ecohydrological principle as a methodology: dual regulation

Based on the first two steps/principles and understanding of their interconnection (Figure 3.4), regulation of hydrological and ecological processes leads us toward low-cost and highly efficient nature-based solutions (Jarosiewicz et al., 2021). In order to fulfill eco-friendliness and low cost, most of the construction materials are recommended to be made of wooden posts which are erected, nailed, and matted by bamboo. To avoid early decomposition of the wooden structure in the soil, different mechanisms of hardening and surface protection can be applied (Figures 3.1–3.3).

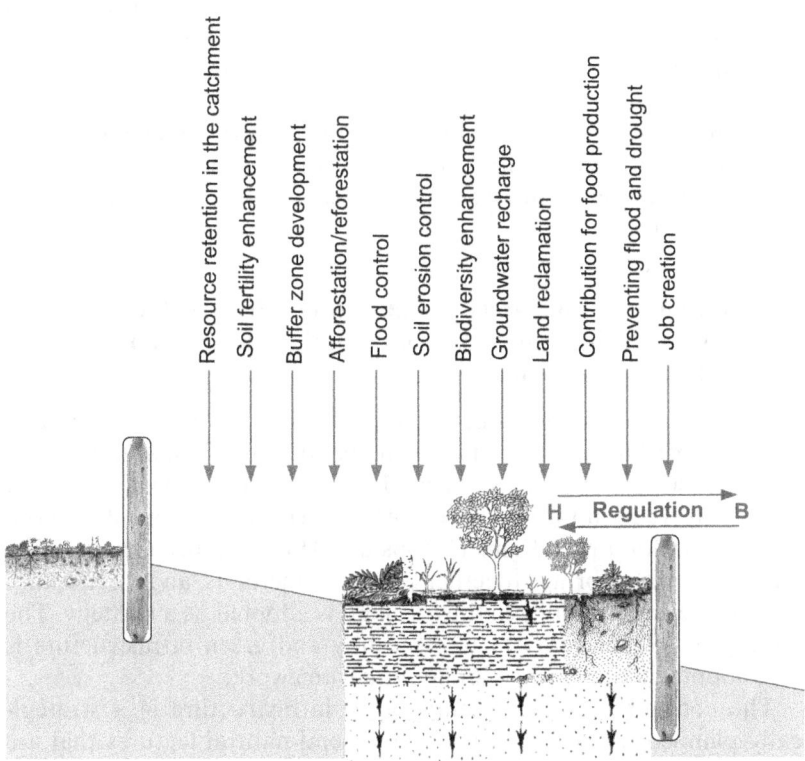

Figure 3.4 Conceptual framework to demonstrate the fulfillment of the third principle of EH by the proposed system.

In general, the proposed system uses the ecosystem properties as environmental management tool which is compliant with the following three assumptions in mind (Zalewski, 2015):

1 *Dual regulation*: The hydrology is purposely regulated using eco-technological interventions as a preamble to the landscape restoration goals followed by biota feedback for ultimate dual regulation that leads to sustainability.
2 *Integration*: The system operates at landscape scale and incorporates various types of biological and hydrological regulation options operating in synergy toward the goal of landscape restoration.
3 *Harmonization:* The proposed ecohydrological system is not a stand-alone intervention. It is harmonized with the prevailing hydrotechnical solutions. It also harmonizes the societal priority needs with the enhanced ecosystem's carrying capacity of the landscape. By harmonization, the plantation action of the system is oriented toward the food security need of the society while fulfilling the ecohydrological dual regulation theory.

The major processes that favor the biota responses include (Figure 3.4): water and nutrients cycling, soil moisture enhancement, sediment trapping, flow energy dissipation, provision of habitat for biota, and erosion control.

3.6 Conception of the green- (semi-) gray infrastructure as an ecohydrological systemic solution (EHSS) for landscape restoration

A system's approach shall be used to identify and solve interrelated problems of a landscape (Watson et al., 1999). In the EcoLaR approach, the ecological restoration of the landscape is set to be the major target of the intervention for it is becoming the central component of modern conservation paradigms (Hobbs and Harris, 2001). On the other hand, EH, which brings together concepts, methods, and disciplinary knowledge from ecology and hydrology is adopted as a strategy. The concept of blended system of semi-gray and green infrastructure is also adopted as an approach/planning framework.

The concept of green- (semi-) gray infrastructure is a strategically planned network of natural and semi-natural features that are designed or managed to deliver a wide range of ecosystem services (Estreguil et al., 2019) and, in our case, it refers to the renovated ecological engineering measures that are able to capture, retain, detain, and slowly release water that might otherwise have passed rapidly along the upstream-downstream continuum of the landscape (Figure 3.5).

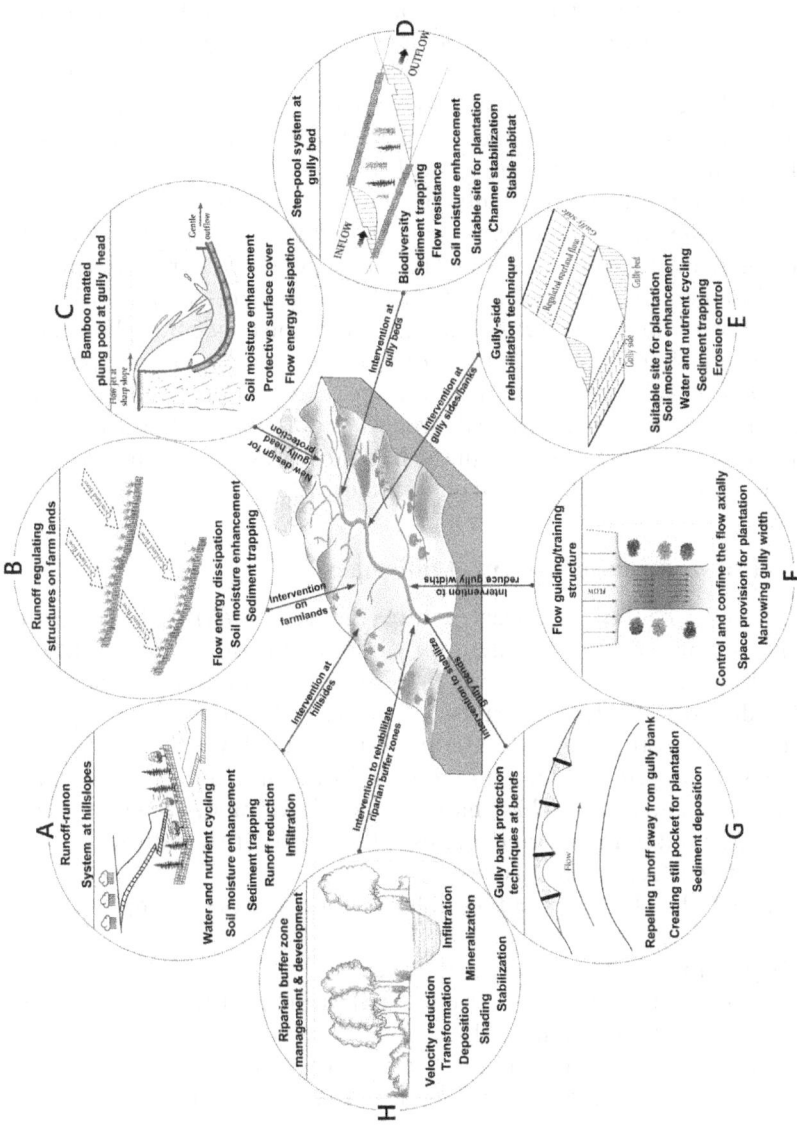

Figure 3.5 Key components of the blended semi-gray and green infrastructure and their ecosystem services.

As shown in Figure 3.5, the proposed ecohydrological systemic solution (EHSS) comprises eight major components that operate in synergy toward ecological restoration of the landscape in a systemic way. The scientific backgrounds of these practices have originated from various disciplines such as landscape ecology (Figure 3.5: #A and #G); soil and water conservation (Figure 3.5: #A, #B, #D, #E, and #H); and hydraulic /river engineering (Figure 3.5: #C, #D, #E, #F, and #G). The basic structural unit (Figure 3.1) is directly applicable for the first five types of physical structures.

The semi-gray components of the infrastructure are meant to regulate the water cycle (*first principle of EH*) along the upstream-downstream continuum of a given landscape. If not harmonized with the 'biota' (*second principle of EH*), these components are just only half way toward landscape restoration and miss the concept of dual regulation (*third principle of EH*). From an ecohydrological perspective, the semi-gray components translate the 'threats' into 'opportunities' through their ecohydrological process regulation mechanisms. The upcoming sub-sections explain the basic features of the major components of the proposed infrastructure in detail.

3.7 Run-off–run-on system at hillslopes

This component of the infrastructure (Figure 3.5: #A) applies the concept of run-off–run-on system for functional restoration of hillsides which in practice may represent degraded forestlands/shrublands or abandoned hill sides. The run-off–run-on system refers to the generation of run-off in one location and infiltrating downslope in an area (Jones et al., 2013). The common environmental problems in such parts of a landscape include moisture deficit for biota growth and unregulated water and nutrient cycles. These areas are usually sources of flood that affect the downstream sites.

Hence, the objectives of the ecohydrologic practices comprise:

- Regulation of water and nutrient cycles
- Enhancement of soil moisture by facilitating infiltration so that the biota growth is enhanced
- Reduction of soil erosion by controlling overland flow
- Retention of sediments and organic matter within the system

The ecohydrologic strategy constitutes creation of a functional system that captures and utilizes vital resources flowing along the overland flow through the process of run-off–run-on (Ludwig et al., 2005). These processes determine the amount of biologically available water

(Falkenmark and Rockstrom, 2004; Newman et al., 2006) for ecohydrologic pulses. The dynamics between these two ecohydrologic units has implications for other ecosystem processes, such as primary productivity, in which the inputs of water and nutrients to resource receiving patches can produce an enhanced pulse of plant growth. The enhanced pulse of plant growth, in turn, should maintain or even increase the capacity of these patches to retain run-off which, in turn, will be translated into increases in per-plant biomass and productivity (Ludwig et al., 2005).

The idea of applying this concept into landscape restoration emerged after critical observation of how landscapes operate to conserve vital resources (water, sediment, nutrients, and organic matters). It acknowledges the importance of increasing the ability of a landscape to retain resources in the landscape (Ludwig et al., 2004) through the concept of landscape patchiness in conserving scarce resources. This, in turn, has important implications for managing the landscapes toward sustainable land use and for the rehabilitation of landscapes that are already degraded (Ludwig and Tongway, 1995).

From the perspective of landscape management, resource retention is an important component of landscape function (Muñoz-Robles et al., 2010), and the type and spatial configuration of run-on and run-off area regulate the redistribution of resources and determine how effectively a landscape can retain resources (Bergkamp, 1998). The general question posed in the formulation of the concept of EH concerns how to regulate the biological processes of freshwater ecosystems using hydrology and – vice versa – how to use biotic ecosystem properties as a tool in water management (Wagner and Zalewski, 2009). There are multiple ways of achieving this linkage (Newman et al., 2006). In our case, the surface run-off is to be captured and stored by the surface restrictions (semi-gray component), a process known as run-off–run-on (Ludwig et al., 2005) which mimics natural functionality of a landscape that can efficiently and effectively conserve and utilize the soil, water, and nutrients (vital resources) within its extent in order to attract and support life. By contrast, landscapes that are dysfunctional tend to lose these vital resources and tend to less likely attract life. Functional integrity of a landscape can be defined as the ability to capture, retain, and use critical resources such as water and nutrients (Ludwig et al., 2004).

In this context, the landscape is functioning as strongly coupled ecological–hydrological systems (Belnap et al., 2004; Seyfried et al., 2004). Some of the pertaining ecohydrological processes in this run-off–run-on system include: water and nutrient cycling, soil moisture enhancement, sediment trapping, run-off reduction, and infiltration.

In the EcoLaR approach, this run-off–run-on system is achieved by establishing a sequence of flow-regulating semi-gray components of the system. The ratio of run-off and run-on length is dictated by the local hydrology as well as the type of farming/biota growth on run-on sites.

3.8 Overland flow-regulating system on farm lands

Overland flow-regulating structure on farm lands such as terraces are built to retain more soil and water, to reduce both hydrological connectivity and erosion through their effect on reducing the slope length, controlling the overland flow and velocity, with positive effects on agricultural activities (Perlotto and D'agostino, 2016). Overland flow in watersheds is responsible for the occurrence of various environmental problems including flood formation, erosion and the transportation of sediments, and the addition of pollutants to the soil (Loewen and Pinheiro, 2017).

The proposed system adopts the concept of terraces with an innovative approach such as: minimum earth work; minimum area loss due to size of the structure; and semi-permeable structure made of bamboo-matted wooden posts (Figure 3.5: #B). By its surface water regulating function, this approach increases the amount of water infiltrating into the soil thereby slowing down and reducing the amount of water leaving the farming system. Some of the prevailing ecohydrologic processes include: flow energy dissipation, soil moisture enhancement, and sediment trapping. The underlying physical law and spacing between two consecutive physical structures are governed by the allowable overland flow length.

3.9 Ecohydrologic strategy to restore gullied landscape

Usually, a gully network contains three parts: head, bed, and banks. In the EcoLaR approach, these locations call for their own ecohydrologic interventions for their restoration as explained below.

3.9.1 Bamboo-matted plunge pool for energy dissipation at gully head

Conventionally, gully heads are treated by reshaping and provision of drop structures. However, application of these techniques to sharp-edged and deep gully heads can be infeasible due to the involvement of significant amount of earth work and further disturbance of the ecosystem. The renovated practice of handling such situation follows the concept of 'plunge pool' of river/hydraulic engineering (Figure 3.5: #C). The combination of protective surface cover (by bamboo mat) and construction of end sills to facilitate water pooling at the gully head convert the waterfall trajectory into gentle outflow. Ecohydrologically, this system helps to enhance soil moisture at the gully head for biota response, provision of protective surface cover, as well as flow energy dissipation.

3.9.2 Step-pool system to regulate run-off at gully beds

The basic intention of soil and water conservation practices at gully beds is to reduce erosive power of the run-off by introducing energy dissipation mechanisms. The common practice of gully bed treatment involves construction of a sequence of check-dams. A check-dam is a small transverse structure designed mainly for three purposes: to control water flow, to conserve soil, and to improve land (Conesa-García and Lenzi, 2010). From an ecohydrological perspective, the self-creating 'step-pool' system of natural streams offers an ecological lesson. A step-pool system is ecologically sound mainly because it provides stable and diversified habitats (Wang et al., 2009), helps in erosion control as well as increasing water surface area (Wang and Yu, 2008), and absorbs considerable amount of energy dissipation through turbulent mixing (Whittaker and Martin, 1982) and also increases the flow resistance, consumes the flow energy and protects the streambed from erosion. The energy-reducing mechanism is primarily the head loss from penetrating the pool system twice – first to get into the pool zone and again to get out. Consequently, the system can be considered as the best ecologically sound pattern in mountainous streams (Wang and Yu, 2008).

In terms of ecohydrologic processes, the proposed step-pool system (Figure 3.5: #D) generally contributes for sediment trapping, creating stable habitat, flow energy reduction, suitable site for plantation, soil moisture enhancement, channel stabilization as well as biodiversity enrichment. The EcoLaR approach adopts a series of flow restriction systems that approximately mimic 'step' and 'pool' system of streams. The basic structural unit of the green- (semi-) gray infrastructure (Figure 3.1) is also applicable here with additional traversal structure to facilitate water pooling along the gully bed.

3.9.3 Gully side rehabilitation

In the conventional system, gully sides are subjected to reshaping and subsequent plantation. Especially in a fragile environment, which is common in degraded landscapes, reshaping practices pose additional instability to the ecosystem. For this reason, the concept of 'run-off–run-on' system is applied in the EcoLaR approach (Figure 3.5: #E) which follows similar features with the proposed ecohydrological practices on hillsides.

3.9.4 Flow guiding/training structure

Streams along their upstream-downstream continuum tend to create flood plains by creating wide portions. One of the objectives of the proposed green infrastructure in EcoLaR approach is creation of habitat for vegetation to growth that, in turn, ecohydrologically regulates the hydrology after its establishment. For this to happen, the wider portions of streams are to be narrowed (Figure 3.5: #F) in order to control and confine the flow axially by guide banks. This practice may result in concentration of flow in the narrowed-section of the gully. Hence, in order to avoid scouring of the gully bed due to the flow concentration, sequence of flow obstruction structure, in our case sequence of the basic structural units of the green- (semi-) gray infrastructure (Figure 3.5), will be installed for subsequent energy dissipation mechanisms.

3.9.5 'Spurs' for gully bank protection at bends

Another characteristic of gullies is their susceptibility to scouring at bends. This situation is neglected by the conventional practices of gully erosion control. The EcoLaR approach, thus, strives to provide mechanisms to deflect the running water away from the gully banks by adopting spurs (sometimes named as spur dykes, groynes, dykes, or transverse dykes) (Figure 3.5: #G) which is one of the river engineering practices. By constructing this structure, the run-off is repelled away from gully bank. Ecohydrologically, the system creates erosion-free zones (still pockets) that favor sediment deposition as well as biota establishment.

3.9.6 Riparian buffer zone management and development

From an ecological and hydrologic perspective, the riparian zone is the area adjacent to a stream that is subject to direct influence of the water in the stream (Coats, 1999). These ecotones are located at the interface between aquatic (*streams, lakes, reservoirs, rivers, and wetlands*) and terrestrial ecosystems (*usually human-disturbed lands*) and link and influence the ecological functioning of both ecosystems (Dosskey et al., 1997; Gregory et al., 1991). A good example of terrestrial ecohydrologic solutions is the creation of the highly efficient land–water transition zones (ecotones) for reduction of pollution fluxes from land to waters from diffused sources (Zalewski, 2014).

Riparian areas have been generally seen as a minor component of the entire landscape system (Wilcox et al., 2017) despite their supporting ecosystem services as habitats for fauna and flora as well as critical providers of ecosystem services to the watershed inhabitants (Soykan and Sabo, 2009). Land/water (riparian) ecotones play a dual role: one is the filtering of nutrients and pollutant transfer along catchment gradients, and the other is flood plain trapping of organic matter, nutrients, and pollutants (Zalewski and Harper, 2001).

Riparian buffers are increasingly portrayed as an important environmental management tool (Parkyn, 2004; Monaghan et al., 2008; Wilcock et al., 2009) and have important implications for water quality and biodiversity (Renouf and Harding, 2015) while also enhancing the physical, chemical, and biological integrity of the terrestrial and aquatic ecosystems. In a landscape context, they are analogous to kidneys because they filter surface and subsurface inputs and reduce sediment and contaminant transport (Pinay et al., 2018).

Buffer zone management usually entails the following four general objectives: erosion control; water quality enhancement; aquatic habitat; and terrestrial habitat improvement (Hawes and Smith, 2005; Lind et al., 2019) in a cost-effective way (Buffler et al., 2005). Conventionally, vegetated buffer strips (VBSs) are established as sole management practices in the riparian buffers which comprise trees, herbs, and grasses (Uusi-Kämppä et al., 2012) (Figure 3.5: #H). Establishment of these layers of vegetative belts usually requires targeted plantation depending on the local condition.

3.10 Decision flow chart for the selection of landscape restoration practices

Landscape restoration strategies usually follow the dichotomy of either passive or active restoration practices. The use of the proposed green- (semi-) gray infrastructure (Figure 3.5) and its components is to be guided by the flow diagram shown on Figure 3.6 below. Once the restoration site is identified, the subsequent steps follow whether the identified management unit needs active of passive restoration actions. The method brings the hillsides (*comprising the landscape component that is found at the very divide of the watershed*) followed by farm lands, gully heads, gully beds, gully sides, gully bends, and riparian zones, respectively.

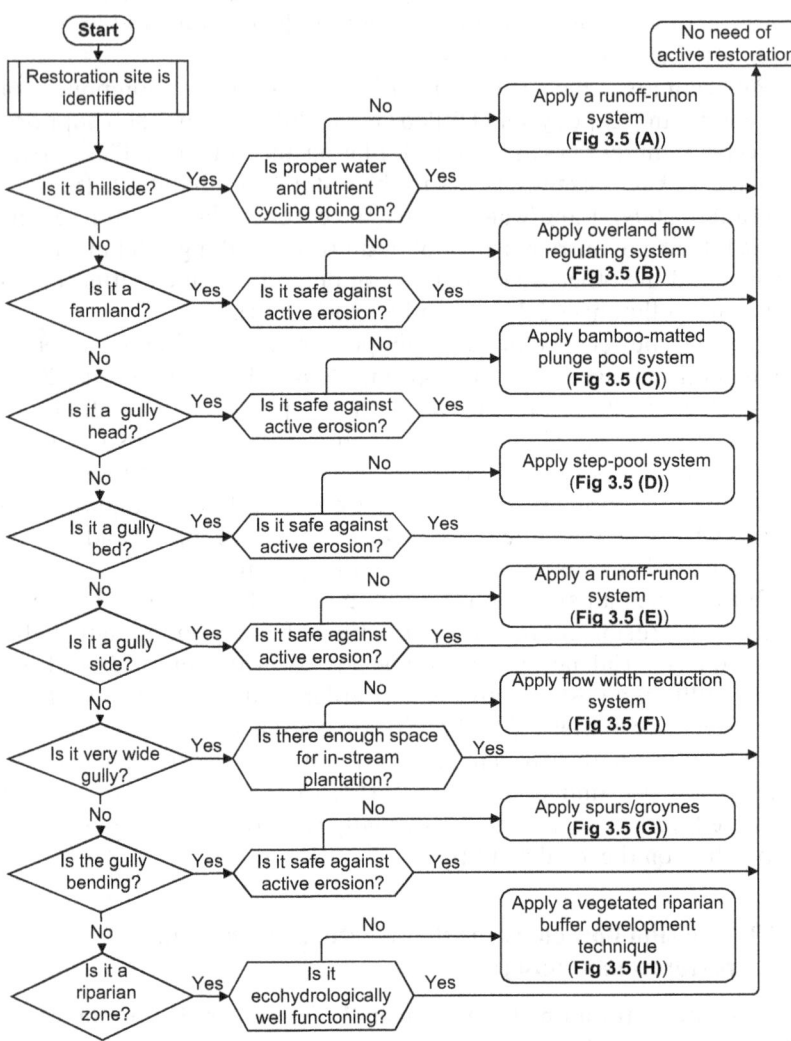

Figure 3.6 Decision support flow chart to guide implementation of the green infrastructure.

3.11 Conclusion

The EcoLaR approach introduces a fairly new way of thinking, seeking solutions, and acting on landscape restoration through managing the water cycle in a holistic manner to achieve the sustainable use of water by societies. However, most of the individual practices comprising the approach have been around us in a fragmented way and sectoral-oriented in practice. The approach brings these individual practices into a system and coined it as a green- (semi-) gray) infrastructure. The synergy among individual techniques of the approach qualifies the three principles of green infrastructure: connectivity, multi-functionality, and spatial planning.

References

Belnap, J., Welter, J.R., Grimm, N.B., Barger, N., & Ludwig, J.A. (2004). Linkages between microbial and hydrologic processes in arid and semiarid watersheds. *Ecology, 86,* 298–307.

Bergen, S.D., Bolton, S.M., & Fridley, J.L. (2001). Design principles for ecological engineering. *Ecological Engineering, 18,* 201–210.

Bergkamp, G. (1998). A hierarchical view of the interactions of runoff and infiltration with vegetation and micro-topography in semiarid shrublands. *Catena, 33,* 201–220.

Bonell, M., Purandara, B.K., Venkatesh, B., Krishnaswamy, J., Acharya, H.A.K., Singh, U.V., Jayakumar, R., & Chappell, N. (2010). The impact of forest use and reforestation on soil hydraulic conductivity in the Western Ghats of India: Implications for surface and sub-surface hydrology. *Journal of Hydrology, 391,* 49–64.

Buffler, S, Johnson, C., Nicholson, J., & Mesner, N. (2005). *Synthesis of design guidelines and experimental data for water quality function in agricultural landscapes in the intermountain west* (USDA Forest Service/UNL Faculty Publications. 13). Lincoln, Nebraska: National Agroforestry Center (NAC).

Carpenter, S.R., Ludwig, D., & Brock, W.A. (1999). Management of eutrophication for lakes subject to potentially irreversible change. *Ecological Applications, 9* (3), 751–771.

Changming, L., Shengtian, Y., Zhiqun, W., Xuelei, W., Yujuan, W., Qian, L., & Haoran, S. (2009). Development of ecohydrological assessment tool and its application. *Science in China Series E: Technological Sciences, 52* (7), 1947–1957.

Chapin, F.S., Matson, P.A., & Mooney, H.A. (2002). *Principles of terrestrial ecosystem ecology.* New York: Springer.

Chazdon, R., & Pedro, B. (2019). Restoring forests as a means to many ends. *Science, 365* (6448), 24–25.

Coats, R.N. (1999). Riparian zone. In R.W. Fairbridge & D.E. Alexander (Eds.), *Environmental Geology.* Dordrecht: Encyclopedia of Earth Science. Kluwer Academic Publishers.

Conesa-García, C., & Lenzi, M.A. (2010). *Check dams, morphological adjustments and erosion control in torrential streams.* Hauppauge, NY: Nova Science Publishers.

Dosskey, M., Schultz, D., & Isenhart, T. (1997). How to design a riparian buffer for agricultural land. Agroforestry Notes (USDA-NAC). *3.*

Eamus, D., Hatton, T., Cook, P., & Colvin, C. (2006). *Ecohydrology-Vegetation function, water and resource management.* Collingwood, Australia: Commonwealth Scientific and Industrial Research Organization.

Estreguil, C., Dige, G., Kleeschulte, S., Carrao, H., Raynal, J., & Teller, A. (2019). *Strategic green infrastructure and ecosystem restoration - Geospatial methods, data and tools.* Joint science for policy report by the joint research centre, the European Environment Agency, the European Topic Centre on Urban, Land, and Soil Systems, and DG Environment.

Falkenmark, M., & Rockstrom, J. (2004). *Balancing water for humans and nature: The new approach in ecohydrology.* London: Earthscan Publications.

Germer, S., Neill, C., Krusche, A.V., & Elsenbeerd, H. (2010). Influence of land-use change on near-surface hydrological processes: Undisturbed forest to pasture. *Journal of Hydrology, 380,* 473–80.

Gregory, S.V., Swanson, F.J., McKee, W.A., & Cummins, K.W. (1991). An ecosystem perspective of riparian zones: Focus on links between land and water. *Bioscience, 41,* 540–551.

Guswa, A.J., Tetzlaff, D., Selker, J.S., Carlyle-Moses, D.E., Boyer, E.W., Bruen, M., Cayuela, C., Creed, I.F., van de Giesen, N., Grasso, D., Hannah, D.M., Hudson, J.E., Hudson, S.A., Iida, S., Jackson, R.B., Katul, G.G., Kumagai, T., Lorens, P., Lopes Ribeiro, F., Michalzik, B., Nanko, K., Oster, C., Pataki, D.E., Peters, C.A., Rinaldo, A., Sanchez Carretero, D., Trifunovic, B., Zalewski, M., Haagsma, M., & Levia, D.F. (2020). Advancing ecohydrology in the 21st century: A convergence of opportunities. *Ecohydrology, 13* (4), e2208

Harper, D., Zalewski, M., & Pacini, N. (2008). *Ecohydrology-processes, models and case studies.* Wallingford: Commonwealth Agricultural Bureau Int. (CABI).

Hawes, E., & Smith, M. (2005). *Riparian buffer zones: Functions and recommended widths.* Yale School of Forestry and Environmental Studies. For the Eightmile River Wild and Scenic Study Committee. New Haven, Connecticut: Yale University.

Hobbs, R., & Harris, J. (2001). Restoration ecology: Repairing the earth's ecosystems in the new millennium. *Restoration Ecology, 9,* 239–246.

IHP (2013). Ecohydrology for Sustainability. Paris: UNESCO.

Jarosiewicz, P., Jurczak, T., & Zalewski, M. (2021). *Ecohydrology for sustainable urban water management.* Pre-Conference for the Second International Conference on 'Water, Megacities and Global Change'. Paris: UNESC.

Jones, O., Gary, S., & Patrick, L. (2013). Using queuing theory to describe steady-state runoff-runon phenomena and connectivity under spatially variable conditions. *Water Resources Research, 49* (11), 7487–7497.

Jørgensen, S.E. (2016). Ecohydrology as an important concept and tool in environmental management. *Ecohydrology & Hydrobiology, 16*(1), 4–6.

Kedziora, A., & Ryszkowski, L. (1999). Does plant cover structure in rural areas modify climate change effects? *Geographia Polonica, 72*(2), 65–88.

Kundzewicz, Z., Saisunee, B., Bronstert, A., Hoff, H., Dennis, L., Lucas, M., & Roland, S. (2002). Coping with variability and change: Floods and droughts. *Natural Resources Forum, 26*, 263–274.

Lind, L., Hasselquist, E.M., & Laudon, H. (2019). Towards ecologically functional riparian zones: A meta-analysis to develop guidelines for protecting ecosystem functions and biodiversity in agricultural landscapes. *Journal of Environmental Management, 249*, 109391.

Loewen, A.R., & Pinheiro, A. (2017). Overland flow generation mechanisms in the Concórdia River basin, in southern Brazil. *RBRH:* Brazilian Journal of Water Resources, *22* (1), e4.

Ludwig, J.A., & Tongway, D.J. (1995). Spatial organization of landscapes and its function in semi-arid woodlands, Australia. *Landscape Ecology, 10*, 51–63.

Ludwig, J.A., Tongway, D.J., Bastin, G.N., & James, C.D. (2004). Monitoring ecological indicators of rangeland functional integrity and their relation to biodiversity at local to regional scales. *Austral Ecology, 29*, 108–120.

Ludwig, J.A., & Tongway, D. (1997). A landscape approach to rangeland ecology. In J.A. Ludwig, D. Tongway, D. Freudenberger, D. Noble, & K. Hodginson (Eds.), *Landscape ecology, function and management: Principles from Australia's Rangelands.* Melbourne: CSIRO Publishing.

Ludwig, J.A., Wilcox, B.P., Breshears, D.D., Tongway, D.J., & Imeson, A.C. (2005). Vegetation patches and runoff-erosion as interacting ecohydrological processes in semiarid landscapes. *Ecology, 86*, 288–297.

Mansourian, S. & Vallauri, D. (2005). *Forest restoration in landscapes: beyond planting trees.* In Mansourian, S. & Vallauri, D. (Eds.). Gland, Switzerland: WWF International.

Monaghan, R.M., de Klein, C.A.M., & Muirhead, R.W. (2008). Prioritisation of farm scale remediation efforts for reducing losses of nutrients and faecal indicator organisms to waterways: A case study of New Zealand dairy farming. *Journal of Environmental Management, 87*, 609–622.

Muñoz-Robles, C., Reid, N., Tighe, M., Briggs, S.V., & Wilson, B. (2010). Soil hydrological and erosional responses in patches and inter-patches in vegetation states in semi-arid Australia. *Geoderma, 560*, 123–134.

Newman, B.D., Wilcox, B.P., Archer, S.R., Breshears, D.D., Dahm, C.N., Duffy, C.J., McDowell, N.G., Phillips, F.M., Scanlon, B.R., & Vivoni, E.R. (2006). Ecohydrology of waterlimited environments: A scientific vision. *Water Resources Research, 42*(6), 1–15.

Parkyn, S. (2004). *Review of riparian buffer zone effectiveness* (MAF Technical Paper No. 2004/05). Wellington Ministry of Agriculture and Forestry. Wellington, New Zealand.

Perlotto, C., & D'agostino, V. (2016). Performance assessment of bench-terraces through 2-D modelling. *Land Degradation & Development, 29* (3), 607–616.

Pinay, G., Bernal, S., Abbott, B.W., Lupon, A., Marti, E., Sabater, F., & Krause. S. (2018). Riparian corridors: A new conceptual framework for assessing nitrogen buffering across biomes. *Frontiers in Environmental Science, 6*(47), 1–11.

Popp, A., Vogel, M., Blaum, N., & Jeltsch, F. (2009). Scaling up ecohydrological processes: Role of surface water flow in water-limited landscapes. *Journal of Geophysical Research Atmospheres, 114*, G04013.

Reid, K.D., Wilcox, B.P., Breshears, D.D., & MacDonald, L. (1999). Runoff and erosion in a pinon-juniper woodland: Influence of vegetation patches. *Soil Science Society of America Journal, 63*, 1869–1879.

Renouf, K., & Harding, J.S. (2015). Characterizing riparian buffer zones of an agriculturally modified landscape. *New Zealand Journal of Marine and Freshwater Research, 49*, 323–332.

Ripl, W. (2003). Water: The bloodstream of the biosphere. *Philosophical Transactions of the Royal Society B: Biological Sciences, 358*, 1921–1934.

SER (Society for Ecological Restoration) (2004). *The SER international primer on ecological restoration.* International Science and Policy Working Group. Tucson/Washington, DC.

Seyfried, M.S., Schwinning, S., Walvoord, M.A., Pockman, W.T., Newman, B.D., Jackson, R.B., & Phillips, F.M. (2004). Ecohydrological control of deep drainage in semiarid regions. *Ecology, 86*, 277–287.

Sola, Phosiso; Judith Oduol; Niguse Hagazi; Sammy Carsan; Rob Kelly; Jonathan Muriuki; Kiros Hadgu; and Maimbo Malesu (2020). *Landscape restoration is more than land restoration: Dryland development in Ethiopia and Kenya.* Wageningen, the Netherlands: Tropenbos International.

Soykan, C.U., & Sabo, J.L. (2009). Spatiotemporal food web dynamics along a desert riparian-upland transition. *Ecography, 32*, 354–368.

Sun, G., Hallema, D., & Asbjornsen, H. (2017). Ecohydrological processes and ecosystem services in the Anthropocene: A review. *Ecological Processes, 6*, 35.

Uusi-Kämppä, J., Turtola, E., Närvänen, A., Jauhiainen, L., & Uusitalo, R. (2012). Phosphorus mitigation during springtime runoff by amendments applied to grassed soil. *Journal of Environmental Quality, 41*, 420–426

Vorosmarty, C.J., & Sahagian, D. (2000). Antropogenic disturbance of the terrestrial water cycle. *Bioscience, 50*, 753–765.

Wagner, I., & Zalewski, M. (2009). Ecohydrology as a basis for the sustainable city strategic planning: Focus on Lodz, Poland. *Reviews in Environmental Science and Bio-Technology, 8*, 209–217.

Wang, L.Q., & Yu, W.D. (2008). Water resources and water environment problems and countermeasures in the Zhangweinan river basin. *Haihe River Conservancy, 5*, 6–8 (in Chinese).

Wang, Z.Y., Melching, C.S., Duan, X.H., & Yu, G.A. (2009). Ecological and hydraulic studies of step-pool systems. *Journal of Hydraulic Engineering, 135*, 705–717.

Watson, C.C., Biedenharn, D.B., & Scott, S.H. (1999). *Channel rehabilitation: Processes, design, and implementation.* Vicksburg, MI: U.S. Army Engineer Research and Development Center.

Whittaker, J.G., & Jaeggi Martin, N.R. (1982). Origin of step-pool system in mountain streams. *Journal of Hydraulic Engineering, 108*(6), 758–773.

Wilcock, R.J., Betteridge, K., Shearman, D., Fowles, C.R., Scarsbrook, M.R., Thorrold, B.S., & Costall, D. (2009). Riparian protection and on-farm best management practices for restoration of a lowland stream in an intensive dairy farming catchment: A case study. New Zealand. *Journal of Marine and Freshwater Research, 43*, 803–818.

Wilcox, B.P., Le Maitre, D.C, Jobbagy, E., Wang, L., & Breshears, D.D. (2017) Ecohydrology: Processes and implications for rangelands. In D. Briske (Ed.), *Rangeland systems*. Springer Series on Environmental Management. Cham: Springer.

Yirdaw E., Tigabu M., & Monge A. (2017). Rehabilitation of degraded dryland ecosystems - review. *Silva Fennica, 51* (1B), 1–32.

Zalewski, M. (2015). Ecohydrology and Hydrologic Engineering: Regulation of Hydrology-Biota Interactions for Sustainability. *Journal of Hydrologic Engineering, 20* (1), A4014012.

Zalewski, M. (2000). Ecohydrology-the scientific background to use ecosystem properties as management tools toward sustainability of water resources. *Ecological Engineering,* 16 (1), 1–8.

Zalewski, M. (2002). Ecohydrology-the use of ecological and hydrological processes for sustainable management of water resources. *Hydrological Sciences Journal, 47* (5), 823–832.

Zalewski, M. (2011). Ecohydrology for implementation of the EU water framework directive. *Water Management, 164* (8), 375–385.

Zalewski, M. (2013). Ecohydrology: Process-oriented thinking towards sustainable river basins. *Ecohydrology & Hydrobiology, 13*(2), 97–103.

Zalewski, M. (2014). Ecohydrology, biotechnology and engineering for cost efficiency in reaching the sustainability of biogeosphere. *Ecohydrology & Hydrobiology, 14*(1), 14–20.

Zalewski, M. (2021). Ecohydrology: An integrative sustainability science. In Theodore T. V. Hromadka II and Prasada P. Rao (Eds.), *Hydrology*. London, UK: IntechOpen.

Zalewski, M., & Harper, D. (2001). Rationale. Ecohydrology-the use of ecosystem properties as management tool for enhancement of absorbing capacity of ecosystem against human impact (UNESCO IHP-V). *Ecohydrology & Hydrobiology, 1*(1–2).

Zalewski, M., & King, C. (2007). Ecohydrology as a management tool. Managing water in diverse ecosystems to ensure human well-being. Hamilton: United Nations University International Network on Water, Environment and Health (UNU-INWEH).

Zalewski, M., & Naiman, R.J. (1985). *The regulation of riverine fish communities by a continuum of abiotic-biotic factors*. In J.S. Alabaster (Ed.). Habitat modification and freshwater fisheries. Butterworths Scientific. London.

Zalewski, M., Janauer, G.A., & Jolankai, G. (1997). *Ecohydrology. A new paradigm for the sustainable use of aquatic resources*. International Hydrological Programme IHP-V Technical Documents in Hydrology.

4 Ecohydrological strategy for functional restoration of landscape hillslopes

Yohannes Zerihun Negussie, Maciej Zalewski, Abebe Beyene Hailu, Mulugeta Dadi Belete, Bekele Beriso Sorsa, Ayualem Ahmed, and Markos Mathewos Godebo

4.1 The underlying concept of the functional restoration of hillslopes through ecohydrologic strategy

Ecohydrologically, hillslopes degradation can be viewed as reduction in the capacity of the land to provide ecosystem goods and services over a period of time for its beneficiaries (Yirdaw et al., 2017); or failure to produce benefits from a particular land use under a specified form of land management (Blaikie and Brookfield, 1987). It involves the reduction of the renewable resource potential due to one or a combination of processes taking place upon the land (FAO, 2011). Degraded landscape is often dysfunctional whereas a non-degraded landscape is typically functional (Alchin, 2011).

Nowadays, landscape restoration efforts, which are powerful mechanisms to recover ecological functionality (César et al., 2021) and long-term socio-ecological processes (Chazdon, 2017), shall address the major environmental challenges of the time that includes land degradation, biodiversity loss, water scarcity, lack of sustainable rural livelihoods, and climate change mitigation and adaptation (Chazdon and Guariguata, 2018). Unsustainable landscape development and use of water bodies often results from the old paradigm-driven management strategies that do not take into account the ecosystem's properties and landscape resilience (Wagner, 2008). Such compelling realities trigger the introduction of ecosystemic approach which underlies the concept of ecohydrologic strategy for landscape restoration.

From a hydrologic perspective, the distribution, growth, and mortality of vegetation on hillslopes are more sensitive to the hydrologic cycle than to any other factor, including nutrients and sunlight

DOI: 10.4324/9781003309130-4

(Weltzin and Tissue, 2003). Once vegetation is degraded, this positive feedback is destabilized, causing a shift to a non-vegetated state (Cook et al., 2009). Ecohydrologic strategy for landscape restoration uses these feedbacks as a management tool in order to enhance quality of ecosystem, which represents a desired endpoint of environmental management (Costanza and Mageau, 1999). According to the first principle of ecohydrology, the hydrological framework serves as a template for functional integration of hydrological and biological processes (Zalewski et al., 1997). If the hydrologic processes are not stable, we do not expect biotic interactions to start to manifest themselves (Zalewski, 2002). We know also that plants need water to survive, and thus, the distribution, composition, and structure of plant communities are directly influenced by spatiotemporal patterns in water availability (Asbjornsen et al., 2011). Furthermore, establishing vegetation on severely degraded landscapes is difficult because of the strong erosive forces (Stokes et al., 2014). In degraded environments, plant growth is often controlled by stochastic pulses of water that directly affect plants' ability to adapt and survive (Schwinning and Sala, 2004). The establishment of vegetation, nevertheless, is possible when combined with engineering structures or ecological engineering techniques (Stokes et al., 2014) that entail design of sustainable ecosystems by integrating human society with its natural environment for the benefit of both (Mitsch and Jorgensen, 2004).

All definitions of ecosystem quality and integrity advocate that soil, water, nutrients, and organic matters (collectively called 'resources'), should not be lost from the system (Noon, 2003). In other words, landscape/hillside functionality is dependent on the conservation and use of soil, water, and nutrient within the landscape/hillside system (Tongway and Ludwig, 2010). In degraded ecosystems, hydrology exerts a profound influence over other abiotic components of the landscape/hillside, primarily erosion (Wainwright et al., 2000) and the loss or redistribution of key plant-limiting nutrients such as nitrogen (Parsons et al., 2003). This knowledge is crucial if we are to effectively address the landscape degradation problem (Wilcox et al., 2003) and creating this understanding to solve these practical problems is the central interest in the field of ecohydrology (Asbjornsen et al., 2011). Ecohydrological solutions will work toward regulating water dynamics and their consequences such as erosion and accelerated nutrient transfer across landscape gradients (Zalewski, 2002).

4.2 The proposed ecohydrologic strategy in the EcoLaR approach

4.2.1 The basic functional unit of the ecohydrologic strategy

Highly functional hillslopes are able to efficiently and effectively conserve and utilize the soil, water, and nutrients (resources) within their extent in order to attract and support life. In contrast, hillslopes that are dysfunctional tend to lose these material resources and tend to less likely attract life. Functional integrity of a landscape can be defined as the ability to capture, retain, and use critical resources such as water and nutrients (Ludwig et al., 2004). This concept is also closely tied to the ability of landscapes to resist stress (stability or resistance) or recover from stress (resilience) (Holling, 1986).

The general question posed in the formulation of the concept of ecohydrology concerns how to regulate the biological processes of freshwater ecosystems using hydrology on the one hand and how to use biotic ecosystem properties as a tool in water management on the other (Wagner, 2008). There are multiple ways in which this linkage can be achieved (Newman et al., 2006). In the EcoLaR (ecohydrology-based landscape restoration) approach, the ecohydrologic strategy for restoration of hillslopes mimics the natural arrangement of resource giving-receiving system (see Figure 4.1) named by different researchers as 'run-off–run-on' system (Ludwig, 2005); 'source-sink' system (Asbjornsen et al., 2011); 'patch-interpatch' system (Ludwig et al., 1997); 'hillslope-patch' system (Tongway and Hindley, 2004); 'divergent and convergent areas'; 'bare slope and vegetated slope' (Belnap et al., 2005); fertile patches and water-shedding interpatches (Alchin, 2011); or 'the source area for generating run-off and the sink area for capturing run-on' (Urgeghe et al., 2010). These two basic units of the landscape are meant to create spatial niches for biota growth. They are designed to function in a way that run-off redistribution from bare to vegetated patches concentrates the critical resources of water, sediment, nutrients, and organic matters which can then enhance vegetation growth and biomass. The ability of the landscape to capture, store, and utilize vital resources (water, nutrients, sediments, and organic matters) is determined by the interactions between these two areas (Alchin, 2011). It is at the hillslope scale that important interactions take place between vegetation patches and run-off (Wilcox et al., 2019).

In practice, the proposed system can be created by a series of physical barriers made of low-cost semi-permeable wooden structures

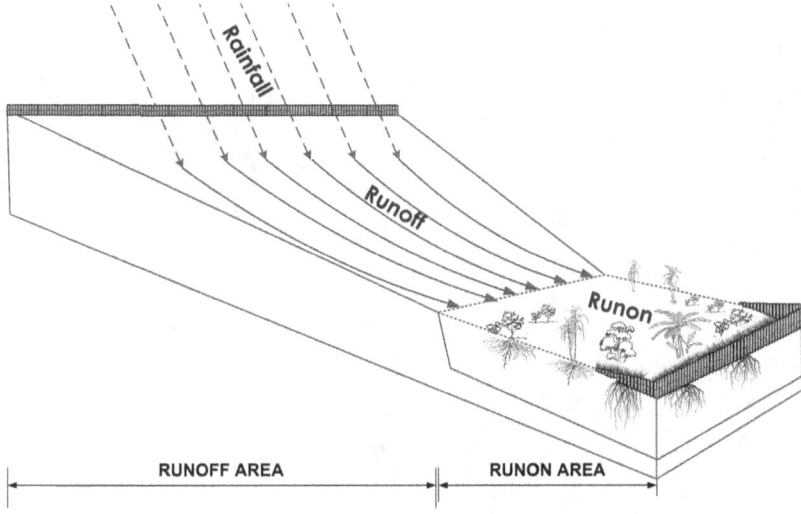

Figure 4.1 Schematics of the proposed resource giving-receiving system at hillslopes.

(with bamboo mat) constructed along the contour and creating the two basic hydrologic units (Figure 4.2). Functionally, the rain falls at the hillslopes and the run-off is redistributed spatially and temporally via a complex assortment of pathways. Such re-concentration of water as run-off from bare patches to run-on patches is important because soil moisture delivered as a deeper pulse is less prone to high evaporative losses and provides more 'plant-available water' (McDonald et al., 2009). The dynamics between these two ecohydrologic units has implications for other ecosystem processes, such as primary productivity, in which the inputs of water and nutrients to resource-receiving patches can produce an enhanced pulse of plant growth. That, in turn, should maintain or even increase the capacity of these patches to retain run-off that will be translated into increases in per-plant biomass and productivity (Ludwig et al., 2005).

The system follows landscape functionality theory, which is based on the recognition that limited water and nutrients cause ecological processes to rely on the interactions of the two spatially nested scales (Alchin, 2011). By landscape function, we are referring to the way the ecological resources are regulated and utilized within a landscape (Sharp, 2011; Tongway and Hindley, 2004; van der Walt et al., 2012). It

Figure 4.2 An example of a gully head with very vertical head in the middle
        of farmlands.
Photograph by the author.

basically recognizes the fact that vegetation growth and productivity
depend not only on the amount of vertical rainfall but also on the
amount of water redistributed laterally by surface run-off–run-on
(Ludwig et al., 2005; Yu et al., 2008). The system also creates spatial
niches for biota growth and it is designed to function in a way that
run-off redistribution from bare to vegetated patches concentrates the
critical resources of water, sediment, nutrients, and organic matters
which can enhance vegetation growth and biomass. The physical barriers that regulate the water dynamics simultaneously facilitate the biota feedback which occupies a key component of the hydrologic cycle
(Asbjornsen, et al., 2011).

The physical structures (the gray component of the system) mimic
a well-organized, resource-conserving natural patch for regulation of
water dynamics. It plays a critical role in the: cycling and redistribution
of vital resources; serving as the site of maximum resource retention,
productivity, and biotic diversity (Kwok et al., 2011); and provision of
stable soil temperature (Pickup, 1985; Pressland and Lehane, 1982). By

these techniques, we are not only 'eliminating threats' of sediment and nutrients transport and flood generation but also 'amplifying the opportunities' by conserving and utilizing these vital and scare resources that would be lost in the absence of the management action. By doing this, we are 'assisting the recovery of the ecosystem that has been degraded, damaged, or destroyed' (SER, 2002).

4.2.2 The approach as an opportunity to derive multiple ecosystem services from an intervention

The ecohydrologic processes that are to be regulated in this hillslope of the landscape include: (1) resource mobilization, (2) transport, (3) accumulation, (4) deposition, (5) cycling, and (6) consumption of critical resources (water, sediment, and nutrients) in the landscape. These ecohydrologic processes are explicitly tied to ecosystem services (Wilcox et al., 2017). As shown in Figure 4.3, some of the ecosystem services derived from the proposed green-(semi-gray) system include water

Figure 4.3 Field trial to demonstrate the incorporation of ecohydrologic strategy in degraded landscape to provide multiple ecosystem services.

storage during the rains and fulfills the functional expectations of water harvesting and conservation measures (*ES 1*) through which climate moderation (*ES 2*); flood control (*ES 3*); and drought management (*ES 4*) benefits are to be obtained that in turn promote groundwater recharge (fluxes toward aquifers) (*ES 5*). Such hydrological regulations will result in erosion control (*ES 6*) and sediment reduction into the lake (*ES 7*) through which water purification (*ES 8*) takes place.

As a result of the above regulating services, the landscape ensures availability of soil moisture which is a physical soil state variable defined as the water contained in the unsaturated soil zone and a key variable in many hydrological, climatological, environmental, and ecohydrological processes. The soil moisture (*green water management*) potentially drives the fixation of dinitrogen (N_2) by microbial symbionts of plants as well as microbial mineralization of soil organic matter. With this, the landscape will be capable of water cycling (*ES 9*) and support nutrient cycling (*ES 10*) by which the landscape provides habitat for fauna and flora (*ES 11*). Through proper delivery of the above services, the ecosystem will provide improved biodiversity and products such as fruits, trees, grasses to the community (*ES 12*) that will also enable the landscape to sequestrate carbon (*ES 13*) and improved soil fertility (*ES 14*) as a result of better nutrient and water cycling. While the above services satisfy environmental and social benefits to the community, the new employment opportunities (*ES 15*) derived from these services contribute to the livelihood and food security aspects of the community, which is a critical aspect for sustainability.

4.3   Field trials to compare the performance of passive vs. active restoration practices at a hillside of Lake Hawassa watershed

4.3.1   Description of the experimental sites and the type of treatments

The trial site is located in the Ethiopian Rift Valley Lakes' Basin (Lake Hawassa sub-basin) (Figure 4.4).

4.3.2   Objectives and specific experimental treatments on the trial sites

Although landscape-scale restoration efforts are gaining more traction worldwide, the success of these restoration actions is unknown in most cases (Watson et al., 2017). This sub-section of the chapter

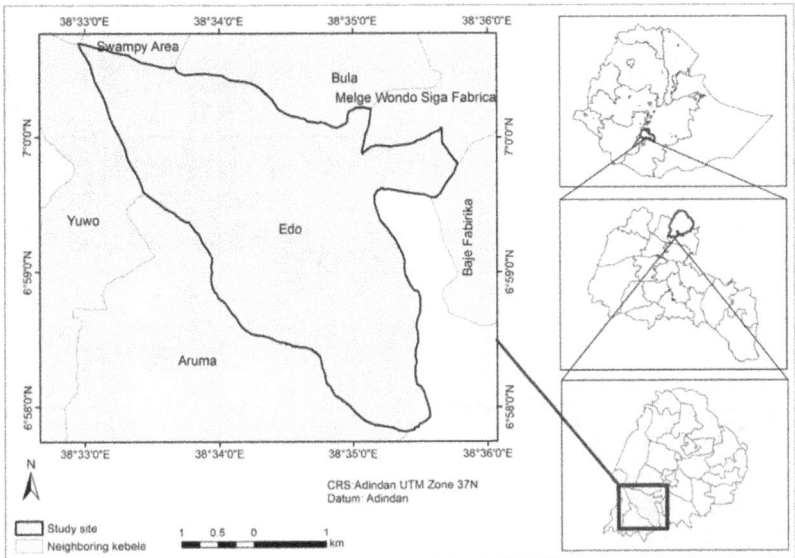

Figure 4.4 Location of the demonstration site.

attempts to compare the performances of 'active' and 'passive' restoration sites (Figure 4.5) in the Ethiopian Rift Valley Basin (Lake Hawassa watershed). The active restoration site receives the proposed ecohydrological system together with area closure (a three-year-old restoration project), while the passive restoration site receives only area closure (protection of the landscape from external pressures such as human and cattle disturbances).

4.3.3 Field layout and data collection strategies

As illustrated in Figure 4.5, the concept of landscape organization was employed to identify resources conserving and leaking portions of the landscape along the general run-off flow direction. Landscape organization is defined as an arrangement of zones that reflect run-on (patch and resource accumulation) and run-off (interpatch and resource loss) processes (Rezaei et al., 2006). In this field trial, the landscape is stratified into discrete units, namely patches and interpatches, to qualify the spatial arrangement of the landscape as recommended by Tongway and Ludwig (2006, 2011). Patches were differentiated as any resource-regulating structure (Tongway & Hindley, 2004) (either

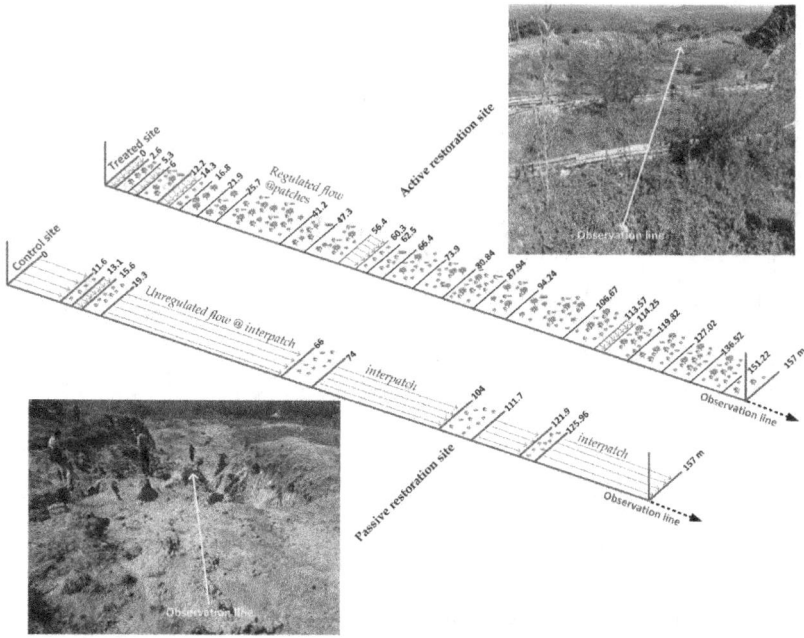

Figure 4.5 Resource giving-receiving system stratifications of the experimental hillslopes as a basis for sample data collection along resource
flowing direction.

engineered or natural); and the spatial arrangement of the patches and
interpatches were identified and measured directly using measuring
tape according to line-intercept criteria (Tongway & Ludwig, 2006).

Technically, patches can be a single plant, a group of plants, rock
or any object that could keep the resources (Kwok et al., 2011; Ludwig
et al., 1999). Their characteristics reflect the functional integrity or the
health of a site (Kwok et al., 2011). In practice, *patches are capturing
and retaining resources such as water, nutrients, and soil particles creating the so-called "fertile zones"* (Ludwig et al., 2005). Interpatches allow the concentration of resources and surface run-off in down slope
patches, resulting in higher biomass production than if resources and
surface run-off were spread uniformly across an area (Ludwig et al.,
2005). They are categorized as a zone where resources are freely transported (English et al., 2005).

Both active and passive sites were cross-sectionally observed with
a total observation line length of 157m. From the figure, it is visible

that there are 22 discrete patches and five interpatches along the 157m transect walk in the active restoration site. On the contrary, only five patches and six interpatches are identified along the passive restoration site with equal lengths of observation lines. The details of patch/interpatch sequences, lengths, widths, and areas are presented in Table 4.1 (for active restoration site) and Table 4.2 (for passive restoration site).

Table 4.1 The landscape organization data for the transect line at treated landscape

Distance (m)	Patch/interpatch sequence as a basis for random sampling	Size of the patch [length (m) × width (m) = Area (m²)]
0 m (start)		
2.6	Interpatch (run-off site)	Low chance of vital resources retention
5.3	Patch (run-on site)	$2.7 \times 7.5 = 20.25$ m²
6	Interpatch (run-off site)	Low chance of vital resources retention
12.2	Patch (run-on site)	$6.2 \times 10.5 = 65.1$ m²
14.3	Interpatch (run-off site)	Low chance of vital resources retention
16.8	Patch (run-on site)	$2.5 \times 10.5 = 26.25$ m²
21.9	Patch (run-on site)	$5.1 \times 15.2 = 77.52$ m²
25.7	Patch (run-on site)	$3.8 \times 25.2 = 95.76$ m²
41.2	Patch (run-on site)	$15.5 \times 28.50 = 441.75$ m²
47.3	Patch (run-on site)	$6.1 \times 28.30 = 172.63$ m²
56.4	Patch (run-on site)	$9.1 \times 30.6 = 278.46$ m²
60.3	Interpatch (run-off site)	Low chance of vital resources retention
62.50	Patch (run-on site)	$2.2 \times 13.9 = 30.58$ m²
66.4	Patch (run-on site)	$3.9 \times 15.2 = 59.28$ m²
73.9	Patch (run-on site)	$7.5 \times 29.3 = 219.75$ m²
80.84	Patch (run-on site)	$6.94 \times 32.6 = 226.24$ m²
87.94	Patch (run-on site)	$7.1 \times 35.2 = 249.92$ m²
94.24	Patch (run-on site)	$6.3 \times 36.9 = 232.47$ m²
106.67	Patch (run-on site)	$12.43 \times 37.3 = 463.64$ m²
113.57	Patch (run-on site)	$6.9 \times 38.4 = 264.96$ m²
114.25	Interpatch (run-off site)	Low chance of vital resources retention
119.82	Patch (run-on site)	$5.57 \times 39.4 = 219.46$ m²
127.02	Patch (run-on site)	$7.2 \times 40.5 = 291.6$ m²
136.52	Patch (run-on site)	$9.5 \times 41.3 = 392.35$ m²
151.22	Patch (run-on site)	$14.7 \times 44.6 = 655.62$ m²
157 m (end)	Patch (run-on site)	$5.78 \times 15.4 = 89$ m²

Table 4.2 The landscape organization data for the transect line at control
landscape

Distance (m)	Patch/interpatch sequence as a basis for random sampling	Size of the patch [length (m) × width (m) = Area (m²)]
0 m (start)		
11.6	Interpatch (run-off site)	Low chance of vital resources retention
13.10	Patch (run-on site)	$1.5 \times 1.6 = 2.4 \, m^2$
15.6	Interpatch (run-off site)	Low chance of vital resources retention
19.3	Patch (run-on site)	$3.7 \times 1.9 = 7.03 \, m^2$
66	Interpatch (run-off site)	Low chance of vital resources retention
74	Patch (run-on site)	$8 \times 21.3 = 170.4 \, m^2$
104	Interpatch (run-off site)	Low chance of vital resources retention
111.7	Patch (run-on site)	$7.7 \times 1.7 = 13.09 \, m^2$
121.9	Interpatch (run-off site)	Low chance of vital resources retention
125.96	Patch (run-on site)	$4.06 \times 1.8 = 7.31 \, m^2$
157 m (end)	Interpatch (run-off site)	Low chance of vital resources retention

4.3.4 Selection of ecohydrological variables and acquisition techniques

Figure 4.6 shows the interdependencies of water, soil, and vegetation
(the three elements of terrestrial ecohydrology) that are presumed to
be influenced by the intervention. Based on this expectation, a total
of 15 environmental variables were employed for this experiment.
Table 4.3 presents the data acquisition methods employed.

4.3.5 Derivation of 'landscape functionality parameters' from field indicators

Landscape function theory and the associated landscape function
analysis (LFA) methodology have become accepted standards for the
ecological assessment of restored environments. This theory can be
used to explain that the ecological functionality of a landscape within
an area can be assessed at a range of spatial scales, ranging from an
individual patch to a whole watershed (Alchin, 2011). Most restoration
programs lack monitoring systems (Machmer and Steeger, 2002). However, assessment of how a landscape can capture and regulate critical

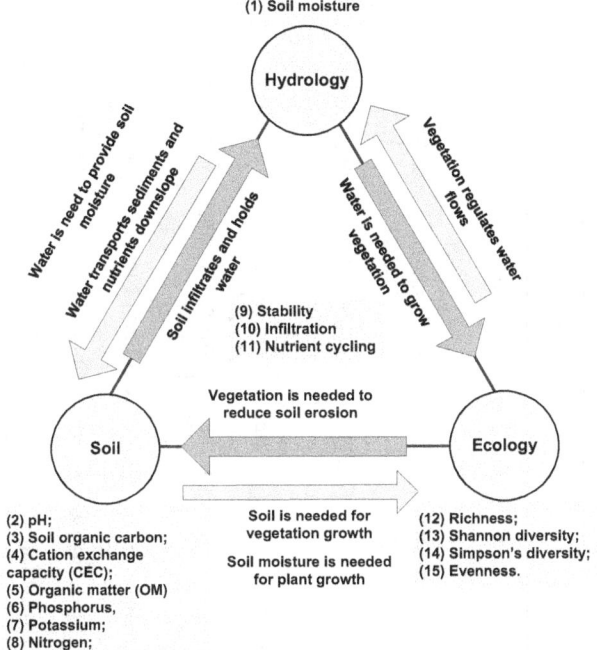

Figure 4.6 Ecohydrological interdependence of hydrology-biota-soil (the three elements of terrestrial ecohydrology).

resources is pivotal to understanding how that landscape is progressing toward self-sufficiency and becoming functional. In parallel, there is a global need for inexpensive tools for research and monitoring and LFA, one of the effective and rapid field research tools (Read et al., 2016), has been used to study ecosystem services provided by systems at different levels of degradation (Valdecantos et al., 2015). The method was developed by Tongway and Hindley (2004) and adopted to measure how successfully the system is functioning biophysically. It is an extremely versatile technique (Green et al., 2009) and has predominantly been used in degraded areas such as mine sites (Van der Walt et al., 2012) and rangelands (Rezaei et al., 2006). It can also be used to quantify changes in soil function response to natural and anthropogenic disturbances such as variation in climate, changes in management practices, and land-use change (Read et al., 2016). Hence, it is becoming increasingly popular worldwide (Kwok et al., 2011) and even sometimes called 'Reading the Landscape' (Jafari footami and Heshmati, 2015).

Table 4.3 The ecohydrological dependent variables with their method of acquisition

	Dependent variables	Ecohydrological implications	Method of data acquisition	References
1	Soil moisture content	To measure the hydrologic regime as a driver for vegetation dynamics and microbial activity	Gravimetric method	Reynolds (1970)
2	Soil pH	To measure of the acidity or basicity of the soil that affect plant growth and biomass yield	Using digital pH meter (1:2.5 soil water ratio)	Edmeades and Wheeler (1990)
3	Soil organic carbon (SOC)	To measure degree of carbon sequestration	Walkley–Black rapid titration method	Walkley and Black (1934)
4	Cation Exchange Capacity (CEC)	To measure the total negative charges within the soil that adsorb plant nutrient cations	By adding 1M ammonium ethanoate (acetate) solution at pH7	Haldar and Sakar (2005)
5	Organic matter (OM)	To measure ecosystem productivity resulting from remains of plants and animals	Calculated by multiplying SOC with Van Bemelen factor	Piper (1950)
6	Total Nitrogen (TN)	To measure degree of soil fertility: for rapid foliage growth and green color	Micro Kjeldhal process	Motsara and Roy (2008)
7	Available Phosphorus (P)	To measure degree of soil fertility: for root formation and growth	Olsen method	Olsen et al. (1954)
8	Available potassium (K)	To measure degree of soil fertility: for overcoming drought stress	By adding 1M ammonium ethanoate (acetate) solution at pH7	Haldar and Sakar (2005)
9	Species richness	To measure the total number of the species in a community	Menhinick's index	Menhinick (1964)

10	Species Diversity	To measure equitability that takes into account both the species richness and evenness dimensions of diversity	Shannon-Wiener Diversity Index (H')	Shannon and Weaver (1949)
11	Species dominance	To measure the number of species present, as well as the abundance of each species	Simpson's index	Simpson (1949)
12	Species evenness or equitability	To measure the relative abundances with which each species is represented in the landscape	Pielou's evenness index	Pielou (1966)
13	Nutrient cycling	To evaluate the way in which elements are continuously being broken down and/ or exchanged for reuse between the living and non-living components of an ecosystem	Landscape functionality analysis (LFA)	Tongway and Hindley (2004)
14	Substrate stability	To compare the degree of substrate stability between the passive and active restoration sites	>>	>>
15	Water infiltration	To compare the degree of water infiltration between the passive and active restoration sites	>>	>>

Landscape functionality analysis categorizes two ecosystem components: (i) landscape organization; and (ii) soil surface condition (Tongway and Hindley, 1995) at relatively fine (<10^4 m^2) scales. These two components reflect the ability of the landscape to capture and retain resources (i.e., the functional integrity). We acknowledge that 'health' is a highly value-laden and context-dependent concept (Tongway and Ludwig, 1997), and in this field trial, the term 'health' is used as a synonym for the functional integrity of the site; thus, the higher the score for the LFA indices, the greater the health of the landscape.

The objective of the derivation procedures was to observe performance of the landscape in terms of the three functional parameters: stability, infiltration, and nutrient cycling (Figure 4.7), which are the surrogates of ecohydrological function and ecosystem function (Maestre and Cortina, 2004). They are strongly associated with the provision and regulation of ecosystem services such as soil retention, cycling of water and nutrients, and carbon storage and biomass production (Read et al., 2016). These parameters were also derived from 11 field indicators (Figure 4.7 and Table 4.4).

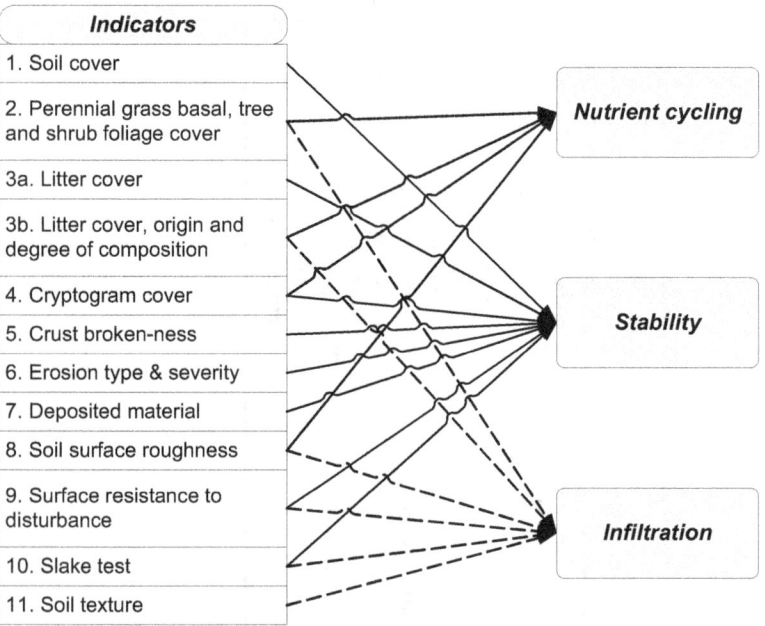

Figure 4.7 Derivation of the main functional parameters from field indicators (modified from Tongway and Hindley, 2004).

Table 4.4 Summary of the 11 indicators of soil biogeochemical properties and processes with their purposes and scoring ranges

	Indicators	Surface feature assessed	Score low-high
1	Rain splash protection/ soil cover	Assess the degree to which the surface cover and projected plant cover ameliorate the effect of raindrops impacting on the soil surface. Assess susceptibility to erosion.	1–5
2	Perennial basal/ vegetation cover	Estimate the basal cover of perennial grass and/or the density of canopy cover of trees and shrubs. Assess the potential biomass for nutrient.	1–4
3	Litter cover	Assess the amount, origin, and degree of decomposition of plant litter. Assess the soil organic matter component and degree of incorporation in the soil.	1–10
4	Cryptogam cover	Assess the cover of cryptogams (algae, fungi, lichens, mosses, liverworts, and mycorrhizas) visible on the soil surface. The presence of cryptogams is a positive indicator of surface stability.	1–4
5	Crust Brokenness	Assess to what extent the surface crust is broken. Assess crust stability and susceptibility to erosion.	1–4
6	Soil erosion type and severity	Assess the type and severity of recent/ current soil erosion.	1–4
7	Deposited materials	Assess the nature and amount of alluvium transported to and deposited on the query zone.	1–4
8	Soil surface roughness	Assess the surface roughness for its capacity to capture and retain mobile resources such as water, propagules, topsoil, and organic matter.	1–5
9	Surface nature	Assess the ease with which the soil can be mechanically disturbed to yield material suitable for erosion by wind or water.	1–5
10	Slake test	Classify the texture of the surface soil and relate this to permeability. Assess the coherence of the soil when it is wet.	1–4
11	Soil texture	Assess the texture class of the surface soil as it affects infiltration.	1–4

The 11 SSA indices were assembled in different combinations (Figure 4.7) by a spreadsheet (Tongway and Hindley, 2004) to calculate three indices of landscape function: (1) surface stability, (2) infiltration capacity, and (3) nutrient cycling potential of the landscape (Tongway and Ludwig, 2006). Thus, every SSA index (scaled from 0 to 100) contributed to the total landscape functionality. Stability (resistance to erosion) refers to the ability of the soil to withstand erosive forces and to reform after disturbance whereas infiltration (capacity for rain and run-on water to infiltrate) is an indicator of the soil's ability to allow water movement into and through the soil profile for root uptake, plant growth and habitat for soil organisms. Moreover, nutrient cycling (organic matter decomposition and cycling) reflects to how efficiently organic matter is cycled back into the soil. Biotic and abiotic processes cause nutrient to flow in and out of a landscape. The nutrient cycling index will be calculated from the aggregate of environmental variables such as perennial grass basal, litter cover, origin and degree of decomposition, cryptogram cover, and surface roughness.

4.3.6 Welch t-test (unequal sample size) for comparison of means between the experimental units

As shown in Figure 4.5 and Table 4.1, the experimental units do not have equal samples. Hence, the test for equality of means is to be done by Welch's *t*-test which is an adaptation of Student's *t*-test (Welch, 1947). This test is more reliable when the two samples have unequal sample sizes (Ruxton, 2006; Derrick et al., 2016) and also it is a robust test for skewed distributions (Fagerland, 2012).

4.4 Results of the experiment

4.4.1 Results of landscape organization analysis

Table 4.5 presents the results of landscape organization analysis. Among these results, the landscape organization index indicates the degree of resource conservation or loss. The active restoration site is found to have a landscape organization index value of 0.94 (= 94% of the gradsect is consisting of resources conserving patches) as compared to 0.16 (= only 16% of the gradsect is consisting of resources conserving patches) in the case of the passive restoration site.

Table 4.5 Results of landscape organization analysis

Landscape organization	Number of patches per 10 m =	Passive restoration	0.32
		Active restoration	1.4
	Total patch area =	Passive restoration	200 m^2
		Active restoration	4,573 m^2
	Landscape organization index* =	Passive restoration	0.16
		Active restoration	0.94
	Average interpatch length =	Passive restoration	22 m
		Active restoration	2 m

* The landscape organization index (LOI) is the proportion of the length of all patches to the total length of the gradsect, thus the percentage of the gradsect that consisted of patches. These parameters reflect the ability of a system to either conserve or lose resources (Tongway & Hindley, 2004). It is calculated as = Length of patches/length of transects. A totally bare transect would have an index of 0 (zero) or if it was all patch) the index would be 1.

4.4.2 Results of soil properties, landscape functionality analysis and biodiversity

As indicated in Table 4.6 and Figure 4.8, the intervention has resulted in significant changes on the five ecohydrological variables (soil moisture, infiltration, stability, nutrient cycling, and soil phosphorus), and less significant effect on the remaining variables.

4.4.3 Results of plant species diversity in the study landscape

Table 4.7 depicts the role of the intervention in the enhancement of biodiversity. In this regard, the active restoration site is shown to increase species richness, diversity, and dominance while it reduces species evenness as influenced by the dominance of few species.

4.5 Conclusion and recommendations

This chapter signifies the promising wider application of the proposed theory and practices of the EcoLaR approach as evidenced by its effectiveness in the context of hillslope as a given landscape. In addition to effectiveness of the approach, it provides an opportunity to view functionality performance of a given landscape from ecohydrologic perspectives.

Table 4.6 Results of soil physiochemical properties; landscape functionality, and biodiversity at passive and active restoration sites

			Samp. 1	Samp. 2	Samp. 3	Samp. 4	Samp. 5	Samp. 6	Samp. 7	Samp. 8	Samp. 9	Samp. 10	Samp. 11	Ave.	St.dev.
Soil moisture and nutrient contents	Moisture content, W (%)	Passive restor.	7.8	5.37	7.38	5.74	7.18	7.86	4.49					6.55	1.33
		Active restor.	11.08	3.66	21.99	14.1	8.58	13.12	7.06	5.82	10.8	14.44	8.3	10.81	5.06
	pH-H$_2$O	Passive restor.	7.67	7.31	7.01	6.75	6.7	6.67	6.75					6.98	0.38
		Active restor.	7.66	7.15	6.85	6.98	7.24	7.48	7.39	6.53	7.14	7.17	7.13	7.16	0.31
	Soil organic carbon (%)	Passive restor.	1.95	3.12	2.86	2.28	1.95	1.81	2.47					2.35	0.5
		Active restor.	2.92	2.6	1.43	2.67	1.76	3.32	2.41	2.28	2.02	2.47	2.6	2.41	0.53
	CEC (Meq/100gm soil)	Passive restor.	11.6	16	35.9	11.2	13	22.4	22.4					18.93	8.84
		Active restor.	16.4	29.8	22.4	29.6	11.3	10.3	8.1	14.4	10.9	13.8	11.6	16.24	7.65
	OM (%)	Passive restor.	3.36	5.38	4.93	3.92	3.65	3.14	4.26					4.09	0.82
		Active restor.	5.04	4.48	2.47	4.59	3.03	5.72	4.15	3.92	3.47	4.26	4.48	4.15	0.91
	Ave.P (PPM)	Passive restor.	2.25	2.22	1.77	1.25	1.12	4.36	5.82					2.68	1.75
		Active restor.	17.38	1.27	14.34	13.91	1.15	5.55	2.85	5.8	3.66	7.81	7.81	7.41	5.55

Parameter	Treatment												Mean	SD
Av.K (mg/Kg soil)	Passive restor.	24	40.9	24.35	26.6	42.75	29.62	36.3					32.07	7.86
	Active restor.	48	46.3	28.5	31	33.85	38.55	38.65	36.85	47.1	42.45	30	38.3	7.02
TN (%)	Passive restor.	0.15	0.25	0.22	0.18	0.14	0.16	0.19					0.18	0.04
	Active restor.	0.23	0.18	0.10	0.21	0.13	0.26	0.19	0.15	0.15	0.19	0.99	0.25	0.25
Landscape functionality — Stability (%)	Passive restor.	47.5	47.5	40	52.5	42.5							46.0	4.9
	Active restor.	82.5	82.5	82.5	80	80	82.5	80	70	72.5	87.5	87.5	80.7	5.4
Infiltration (%)	Passive restor.	14.04	14.04	14.04	25.93	14.04							16.4	5.3
	Active restor.	51	44.9	44.9	48.4	46.6	47.5	45.7	38.7	49.3	47.5	38.7	45.7	3.9
Nutrient cycling (%)	Passive restor.	11.6	16.3	9.3	38.5	11.6							17.5	12.0
	Active restor.	47.7	37.2	41.9	41.9	39.5	45.3	43	29.1	31.4	43	43	40.3	5.7

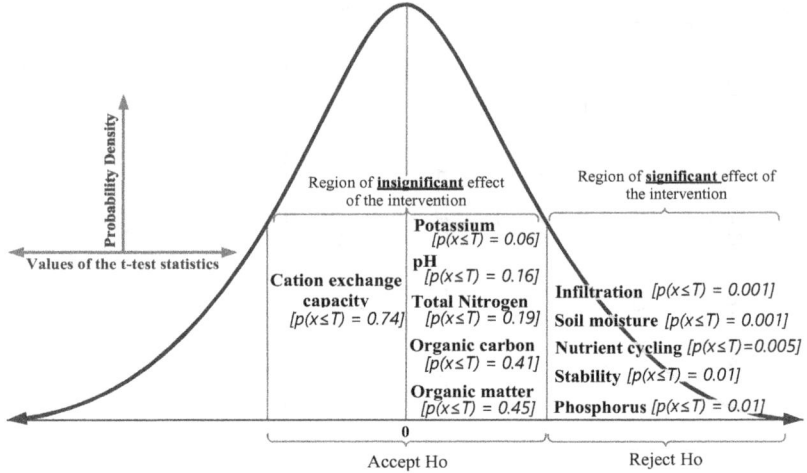

Figure 4.8 Schematic presentation of *t*-test results.

Table 4.7 Results of species diversity

Biodiversity	Richness	Passive restoration	8
		Active restoration	18
	Shannon diversity (H')	Passive restoration	0.44
		Active restoration	1.22
	Simpson's diversity (D)	Passive restoration	0.28
	(*dominance*)	Active restoration	0.62
	Evenness (H'/Hmax)	Passive restoration	0.99
		Active restoration	0.96

References

Alchin, M.D. (2011). *A test of landscape function theory in the semi-arid shrub-lands of Western Australia* (Doctoral dissertation, Curtin University, Department of Environment and Agriculture – School of Science). Retrieved from https://espace.curtin.edu.au/handle/20.500.11937/1498.

Asbjornsen, H., Goldsmith, G.R., Alvarado-Barrientos, M.S., Rebel, K., Van Osch, F.P., Rietkerk, M., Chen, J., Gotsch, S., Tobón, C., Geissert, D.R., Gómez-Tagle, A., Vache, K., & Dawson, T.E. (2011). Ecohydrological advances and applications in plant–water relations research: A review. *Journal of Plant Ecology, 4*(1–2), 3–22.

Belnap, J., Welter, J., Grimm, N., Barger, N., & Ludwig, J. (2005). Linkages between microbial and hydrologic processes in arid and semiarid watersheds. *Ecology, 86*, 298–307.

Blaikie, P., & Brookfield, H. (1987). *Land Degradation and Society*. London & New York: Methuen.

César, R.G., Belei, L., Badari, C.G., Viani, R.A.G., Gutierrez, V., Chazdon, R.L., Brancalion, P.H.S., & Morsello, C. (2021). Forest and landscape restoration: A review emphasizing principles, concepts, and practices. *Land, 10*(1), 28.

Chazdon, R. (2017). Landscape restoration, natural regeneration, and the forests of the future. *Annals of the Missouri Botanical Garden, 102*, 251–257.

Chazdon, R.L. & Guariguata, M.R. (2018). *Decision support tools for forest landscape restoration: Current status and future outlook* (Occasional Paper 183). Bogor: CIFOR.

Cook, B.I., Miller, R.L., & Seager, R. (2009). Amplification of the North American "Dust Bowl" drought through human-induced land degradation. *Proceedings of the National Academy of Sciences, 106*(13), 4997–5001.

Costanza, R., & Mageau, M. (1999). What is a healthy ecosystem? *Aquatic Ecology, 33*(1), 105–115.

Derrick, B., Toher, D., & White, P. (2016). Why Welchs test is Type I error robust. *The Quantitative Methods for Psychology, 12*(1), 30–38.

Edmeades, D.C., & Wheeler, D.M. (1990). Measurement of pH in New Zealand soils: An examination of the effect of electrolyte, electrolyte strength, and soil:solution ratio. *New Zealand Journal of Agricultural Research, 33*(1), 105–109.

English, P., Wallbrink, P., Humphreys, G., Shakesby, R., Doerr, S., Blake, W., Chafer, C., & Vigneswaran, B. (2005). *Impacts on water quality by sediments and nutrients released during extreme bushfires*. Nattai National Park, NSW: Report 2: Tracer assessment of post-fire sediment and nutrient redistribution on hillslopes, Sydney Catchment Authority - CSIRO Land & Water Collaborative Research Project-CSIRO Land and Water Client Report.

Fagerland, M.W. (2012). T-tests, non-parametric tests, and large studies - A paradox of statistical practice? *BMC Medical Research Methodology, 12*, 78–85.

FAO (2011). *The state of the world's land and water resources for food and agriculture (SOLAW) - Managing systems at risk*. Rome and Earthscan, London: Food and Agriculture Organization of the United Nations.

Green, E., Mitchell, B., Tongway, D., Doherty, M., Beaty, M., & Brack, C. (2009). Measuring and monitoring urban ecological function in Canberra. Report 2: Methods for the measurement of urban ecosystem function and implications for managing the ANU campus: a pilot study. Report Number: USP2008/013 (CAF R-555-14). CSIRO ecosystem sciences, Canberra.

Haldar, A., & Sakar, D. (2005). *Physical and chemical method in soil analysis: Fundamental concepts of analytical chemistry and instrumental techniques*. New Delhi: New Age International (P) Ltd. Publisher.

Holling, C.S. (1986). The resilience of terrestrial ecosystems: Local surprise and global change. In W.C. Clark and R. Munn (Eds.), *Sustainable development of the biosphere*. Cambridge: Cambridge University Press.

Jafari footami, I., & Heshmati, G. (2015). A Comparison of the performance of LFA method with Traditional assessment methods of soil properties in summer rangeland ecosystems, Hezar Jerib, North of Iran. *Journal of Soil Environment 1*, 28–34.

Kwok, A.B.C., Eldridge, D.J., & Oliver, I.A.N. (2011). Do landscape health indices reflect arthropod biodiversity status in the eucalypt woodlands of eastern Australia? *Austral Ecology, 36*, 800–813.

Ludwig, J.A., Tongway, D.J., Bastin, G.N., & James, C.D. (2004). Monitoring ecological indicators of rangeland functional integrity and their relation to biodiversity at local to regional scales. *Austral Ecology, 29*, 108–120.

Ludwig, J.A., Tongway, D.J., Eager, R.W., Williams, R.J., & Cook, G.D. (1999). Fine-scale vegetation patches decline in size and cover with increasing rainfall in Australian savannas. *Landscape Ecology, 14*, 557–566.

Ludwig, J.A., Wilcox, B.P., Breshears, D.D., Tongway, D.J., & Imeson, A.C. (2005). Vegetation patches and runoff-erosion as interacting ecohydrological processes in semiarid landscapes. *Ecology, 86*, 288–297.

Ludwig, J., Tongway, D., Freudenberger, D., Noble, J., & Hodgkinson, K. (1997). *Landscape ecology, function and management: Principles from Australia's Rangelands*. Melbourne: CSIRO Publishing.

Machmer, M. & Steeger, C. (2002). *Effectiveness evaluation guidelines for ecosystem restoration*. Victoria: Final report submitted to Habitat Branch, Ministry of Water, Land and Air Protection.

Maestre, F.T. & Cortinal, J. (2004). Insights into ecosystem composition and function in a sequence of degraded semiarid steppes. *Restoration Ecology, 12*(4), 494–502

McDonald, A.K., Kinucan, R.J., & Loomis, L.E. (2009). Ecohydrological interactions within banded vegetation in the northern Chihuahuan Desert, USA. *Ecohydrology, 2*, 66–71.

Menhinick, E.F. (1964). A comparison of some species-individuals diversity indices applied to samples of field insects. *Ecology, 45*, 859–861.

Mitsch, W.J., & Jorgensen, S.E. (2004). *Ecological engineering and ecosystem restoration*. New York: John Wiley & Sons.

Motsara, M.R. & Roy, R.N. (2008). Guide to laboratory establishment for plant nutrient analysis. Food and Agriculture Organization of the United Nations, Rome.

Newman, B.D., Wilcox, B.P., Steven, R. A., Breshears, D.D., Dahm, C.N., Christopher J. Duffy, C.J., Nathan G. McDowell, N.G., Phillip, F.M., Bridget R. Scanlon, B.R., & Vivoni, E.R. (2006). Ecohydrology of water-limited environments: A scientific vision. *Water Resources Research, 42*(6), W06302

Noon, B.R. (2003). Conceptual issues in monitoring ecological resources. In D.E. Busch & J.C.Trexler (Eds), *Monitoring ecosystems: Interdisciplinary approaches for evaluating ecoregional initiatives*. Washington, DC: Island Press.

Olsen, S.R., Cole, C.V., Watanabe, F.S., & Dean, L.A. (1954). *Estimation of available phosphorus in soils by extraction with sodium bicarbonate* (Circular, 939). Washington, DC: US Department of Agriculture.

Parsons, A.J., Wainwright, J., Schlesinger, W.H., & Abrahams, A.D. (2003). The role of overland flow in sediment and nitrogen budgets of mesquite dunefields, southern New Mexico. *Journal of Arid Environments, 53*, 61–71.

Pickup, G. (1985). The erosion cell - A geomorphic approach to landscape classification in range assessment, *Australia Rangeland Journal, 7*(2), 114–121.

Pielou, E.C. (1966). The measurement of diversity in different types of biological collection. *Journal of theoretical biology, 13*, 131–144.

Piper, C.S. (1950). Soil and plant analysis. Adelaide: The University of Adelaide Press.

Pressland, A., & Lehane, K. (1982). Runoff and the ameliorating effect of plant cover in the mulga communities of south western Queensland. *Australian Rangeland Journal, 4*(1), 16–20.

Read, Z.J., King, H.P., Tongway, D.J., Ogilvy, S., Greene, R.S.B., & Hand, G. (2016). Landscape function analysis to assess soil processes on farms following ecological restoration and changes in grazing management. *European Journal of Soil Sciences, 67*(4), 409-420.

Reynolds, S.D. (1970). The gravimetric method of soil moisture determination Part III An examination of factors influencing soil moisture variability. *Journal of Hydrology, 11*(3), 288–300.

Rezaei, S.A., Arzan, H., & Tongway, D. (2006). Assessing rangeland capability in Iran using landscape function indices based on soil surface attributes. *Journal of Arid Environments, 65*, 460–473.

Ruxton, G.D. (2006). The unequal variance t-test is an underused alternative to Student's t-test and the Mann--Whitney U test. *Behavioral Ecology, 17*(4), 688–690.

Schwinning, S., & Sala, O.E. (2004). Hierarchy of responses to resource pulses in and semi-arid ecosystems. *Oecologia, 141*, 211–20.

SER (Society for Ecological Restoration Science and Policy Working Group) (2002). The SER Primer on Ecological Restoration. Retrieved from http://www.ser.org/.

Shannon, C.E., & Weaver, W. (1949). *The mathematical theory of communication.* Urbana, IL: University of Illinois Press.

Sharp, S.B. (2011). *Landscape function in canberra nature park and impacts of threatening processes on landscape function.* Canberra: Report to the commissioner for sustainability and the environment.

Simpson, E.H. (1949). Measurement of diversity. *Nature, 163*, 688.

Stokes, A., Douglas, G.B., & Fourcaud, T. (2014). Ecological mitigation of hillslope instability: Ten key issues facing researchers and practitioners. *Plant Soil, 377*, 1–23

Tongway, D.J., & Hindley, N. (1995). *Assessment of soil condition of tropical grasslands.* Canberra: CSIRO Ecology and Wildlife.

Tongway, D.J., & Ludwig, J.A. (2010). *Restoring disturbed landscapes: Putting principles into Practice.* Washington, DC: Island Press.

Tongway, D.J., & Ludwig, J.A. (1997). The conservation of water and nutrients within landscapes. In J.A. Ludwig, D. Tongway, D. Freudenberger, J.

Noble, & K. Hodgkinson (Eds.), *Landscape ecology function and management: Principles from Australia's rangelands.* Melbourne: CSIRO.

Tongway, D.J., & Hindley, N. (2004). Landscape function analysis: A system for monitoring rangeland function. *African Journal of Range & Forage Science, 21*(2), 109–113.

Tongway, D.J., & Ludwig, J.A. (2006). Resource retention and ecological function as restoration targets in semi-arid Australia. *Restoration Ecology, 14,* 369–378.

Tongway, D.J., & Ludwig, J.A. (2011). *Restoring disturbed landscapes: Putting principles into practice.* Washington, Covelo, London: Island Press.

Urgeghe, A.M., Breshears, D.D., Martens, S.N., & Beeson, P.C. (2010). Redistribution of runoff among vegetation patch types: On ecohydrological optimality of herbaceous capture of run-on. *Rangeland Ecology Manage, 63,* 497–504.

Valdecantos, A., Vallejo, V., Hessel, R., Ritsema, C.J., & Ritsema, C.J. (2015). *Report on structural and functional changes associated to regime shifts in Mediterranean dryland ecosystems* (CASCADE Project Deliverable 5.1). Centra de Estudios Ambientales del Mediterr á neo (CEAM), Alicante, Spain.

van der Walt, L., Cilliers, S. S., Kellner, K., Tongway, D., & Van Rensburg, L. (2012). Landscape functionality of plant communities in the impala platinum mining area, Rustenburg. *Journal of Environmental Management, 113,* 103–16.

Wagner, I. (2008). *Understanding ecohydrological processes for sustainable floodplain management.* International Institute of Polish Academy of Science. Lodz, Poland.

Wagner, J.R. (2008). Landscape aesthetics, water, and settler colonialism in the Okanagan valley of British Columbia. *Journal of Ecological Anthropology, 12*(1), 22–38.

Wainwright, J., Parsons, A.J., & Abrahams, A.D. (2000). Plot-scale studies of vegetation, overland flow and erosion interactions: Case studies from Arizona and New Mexico. *Hydrological Processes, 14,* 2921–2943.

Walkley, A. & Black, I.A. (1934). An examination of the Degtjareff Method for determining soil organic matter, and a proposed modification of the chromic acid titration method. *Soil Science, 37*(1), 29–38

Watson, D.M., Veronica, A.J.D., Banks, S.C., Driscoll, D.A., van der Reed, R., Doerr, E.D., & Sunnucks, P. (2017). Monitoring ecological consequences of efforts to restore landscape-scale connectivity. *Biological Conservation, 206,* 201–209.

Welch, B.L. (1947). The generalization of "Student's" problem when several different population variances are involved. *Biometrika, 34*(1–2), 28–35.

Weltzin, J.F., & Tissue, D.T. (2003). Resource pulses in arid environments-Patterns of rain, patterns of life. *New Phytologist, 157,* 171–173.

Wilcox B.A., Aguirre, A.A., Padua, N.D., Siriaroonrat, B., & Echaubard, P. (2019). Operationalizing one heath employing social-ecological systems theory: Lessons from the Greater Mekong Subregion. *Frontiers in Public Health, 7* (85), 1–12.

Wilcox B.P., Le Maitre D.L., Jobbagy, E., Wang, L., & Breshears, D.D. (2017). Ecohydrology: Processes and implications for rangelands. In D. Briske (Ed.), *Rangeland systems*. Springer Series on Environmental Management. Cham: Springer.

Wilcox, B.P., Breshears, D.D., & Allen, C.D. (2003). Ecohydrology of a resource conserving semiarid woodland: Effects of scale and disturbance. *Ecological Monographs, 73*, 223–239.

Yirdaw, E., Tigabu, M., & Monge, A. (2017). Rehabilitation of degraded dryland ecosystems - Review. *Silva Fennica, 51*(1B), 1673.

Yu, M., Gao, Q., Epstein, H.E., & Zhang, X. (2008). An ecohydrological analysis for optimal use of redistributed water among vegetation patches. *Ecological Applications, 18*, 1679–1688.

Zalewski, M. (2002). Ecohydrology, the use of ecological and hydrological processes for sustainable management of water resources. *Hydrological Sciences-Journal des Sciences hydrologiques, 47*(5), 823–832.

Zalewski, M., Janauer, G.A., & Jolankai, G. (1997). *Ecohydrology: A new paradigm for the sustainable use of aquatic resources* (Technical documents in hydrology No.7. UNESCO, IHP). Paris: International Hydrological Programme.

5 Ecohydrologic strategy for sustainable management of farmlands

Mulugeta Dadi Belete

5.1 Introduction

Farmlands are one of the elements of a landscape that cover more than one-third of the Earth's land surface and are perhaps our most vital ecosystems to sustain humankind (UN, 2021) that are most directly managed by people to meet human goals such as food, fiber, and fuel needs (Swinton et al., 2007; Wood et al., 2001). This ecosystem can supply all the three major categories of ecosystem services – provisioning, regulating, and cultural services if their demand of supporting services that enable it to be productive is satisfied (Swinton et al., 2007). It is both providers and consumers of ecosystem services in which the nutrient cycling and the provision of water are among the major supporting services. Their regulating services may also include flood control, water quality control, carbon storage and climate regulation through greenhouse gas emissions, disease regulation, and waste treatment (Power, 2010).

In the setting of agricultural ecosystem, sustainable land management (SLM) is an instrumental approach that aims at harmonizing the complimentary goals of providing environmental, economic, and social opportunities for the benefit of present and future generations while maintaining and enhancing the quality of the land (soil, water and air) resources (Smyth and Dumanski, 1993). Effective SLM properly combines technologies, policies, and activities to realize sustainability that is characterized by productivity, stability/resilience, protection, viability, and acceptability/equity (Smyth and Dumanski, 1993). It is generally recognized that ecologically balanced land management can achieve both economic and environmental benefits and serve as a foundation (linch pin) for further rural interventions (investments) (World Bank, 1997). Despite the socioeconomic dimension of SLM approach, some of the prevalent technologies such

DOI: 10.4324/9781003309130-5

as soil bund and fanya juu unnecessarily consume significant amount of productive lands ranging from 8% (Tenge et al., 2005) up to 30% (Teshome et al., 2013) of farmers' holdings. This situation evidently affects the adoption rate for it is economically less motivational in the eyes of the farmers (Belete, 2021). Prevalence of this reality contradicts with the basic essence of SLM that advocates optimal use of land resources and social acceptability of the actions that eventually affect sustainability of the farming system. Thus, shifting into an alternative approach that is more ecological is very timely.

By definition, SLM is the adoption of land use systems that enables land users to maximize the economic and social benefits from the land while maintaining or enhancing the ecological support functions of the land resources (FAO, 2009) within the context of the potentially devastating effects of climate change. This notion of SLM entails the consideration of both hydrology and ecology *ecohydrology*) to sustainably manage land resources. The upcoming sub-section elaborates on a number of common features shared between ecohydrology (*more of terrestrial phase*) and SLM concepts. It is found that both concepts implicitly:

1 Consider cycling of water and nutrients in a landscape
2 Strive to harmonize their specific solutions of environmental management with societal needs
3 Entertain the threats of climate change
4 Consider the catchment/landscape as an ecosystem where hierarchy of factors are operational
5 Follow the concept of Abiotic-Biotic Regulatory Continuum (ABRC) that assumes stability of the abiotic factors as a primarily important factor for the positive manifestation of biotic feedback (Zalewski and Naiman, 1985)
6 Attempt to increase the 'carrying capacity' (water quality, restoration of biodiversity, ecosystem services for society, and resilience) of a landscape
7 Advocate the use of ecological engineering solutions while acting on the land
8 Rely on the beneficial interaction among soil-plant-water-biodiversity to solve environmental problems
9 Target to accomplish multiple services
10 Integrate various types of biological and hydrological regulations toward achieving synergy to improve quality of the environment
11 Highlight water as a key driver of ecosystem dynamics

In line with these common features, this chapter attempts to integrate the strategy of terrestrial ecohydrology into SLM for the betterment of the conventional environmental management efforts.

5.2 Surface runoff (overland flow) as a key hydrologic parameter for sustainable land management

The single greatest human activity affecting water flows in landscapes is agriculture (Figure 5.1). Therefore, ensuring efficient use of land and water in farmlands is of paramount importance (Falkenmark, 2016).

In the context of ecohydrology-based landscape restoration (EcoLaR) approach, regulation of the surface runoff (overland flow) on agricultural lands is considered as a key design parameter in renovating the conventional practices of land management. An overland flow is the type of flow that runs off downslopes and first forms thin sheets of water and covers nearly all the ground in what is termed as

Figure 5.1 Untreated farmland subjected to unregulated runoff causing significant erosion.
Photograph by the author.

sheet flow (Shroder and Ahmadzai, 2016). It is infiltration excess and initial phase of surface runoff (Emmett, 1970) and develops after water ponded on the land surface rises high enough that it can begin to flow downslope (Kampf and Mirus, 2013) and potentially produces sheet erosion. This approach ensures that the amount of water seeping into the soil is maximized thereby slowing down and reducing the amount of water running off (Sustainet, 2010).

Overland flow in landscapes is responsible for the occurrence of various environmental problems including flood formation, erosion, the transportation of sediment, and the addition of pollutants to the soil (Loewen and Pinheiro, 2017). It is part of the rainfall that flows off over the surface of the land in non-concentrated form or in temporary and very shallow channels. Closely related to the hydraulic properties of overland flow is the ability of these extremely shallow flows to re-work the ground surface over which they flow. Such reworking occurs in nature and is evidenced by the sheet erosion of hillslope sediments and the transportation of these sediments, which are either deposited at some other location on the slope or are carried entirely out of the drainage system (Emmett, 1970).

When Hortonian overland flow concentrates on hillslopes due to micro- or macro-topography, shear stress generated by flowing water can exceed the shear strength of the soil, causing rills and progressively gullies to form (Sidle et al., 2019). Hydraulically, the major objective of farmland structures such as soil bunds and fanya juu terraces is to regulate the overflow before it attains erosive power. It is recommended that all approaches to soil conservation, i.e., agronomic, soil management, and mechanical means, be used to manage runoff from the land (Reinders et al., 2016). In the EcoLaR approach, this objective is linked to the first principle of ecohydrology (*regulating the hydrology*) that should be achieved through ecological engineering options in order to avoid wastage of productive lands, to minimize earth works, as well as to avoid over-engineering of the agricultural ecosystem. The upcoming sub-section describes the proposed techniques of restoring functionality of agricultural ecosystems.

5.3 Simple eco-technological cross-slope barriers to regulate surface hydrology in farmlands

In the context of EcoLaR approach, an overland flow regulation system made of cost-efficient local materials (Figures 5.2 and 5.3) is

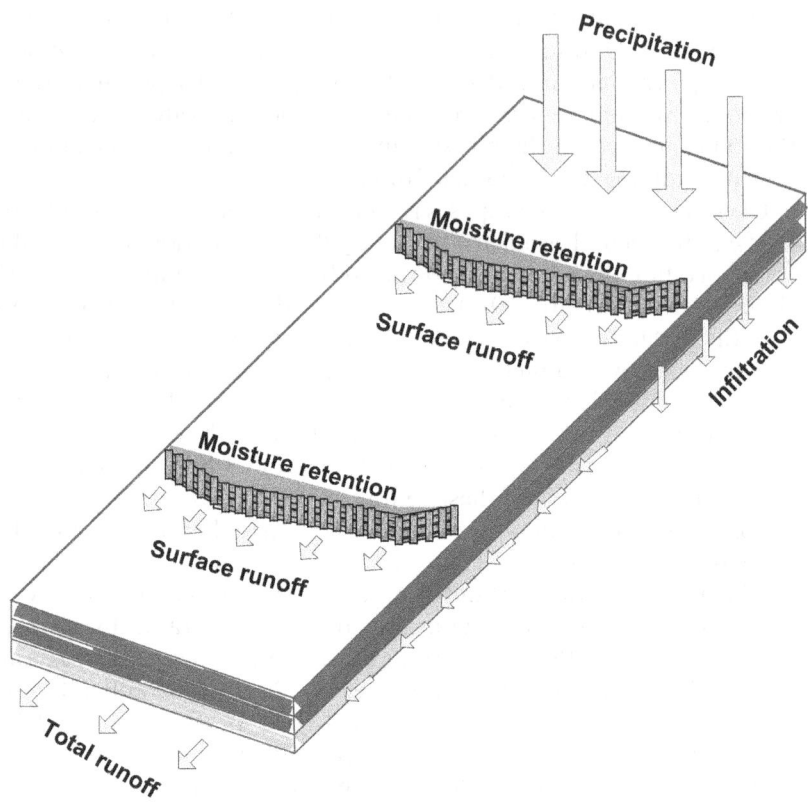

Figure 5.2 Typical schematics of the structure in agricultural lands.

apparently renovated as part of the SLM practice. The renovated system basically comprises two elements: the physical structure and the vegetation element. At an early stage of its establishment, the physical structure regulates the surface flow with more or less similar function as the conventional terracing practices. In the perspectives of ecohydrology, the function of the physical structure aligns with the first principle of ecohydrology. Once the physical structure is in place, the vegetational element follows. For its complete ecohydrological role, place-based and need-driven plantation of grasses, fruits, trees, shrubs, etc. shall be planted along the physical structure for at least two reasons: (1) for their 'green feedback'– as a stabilizing factor for solar energy, water dynamics, and biogeochemical cycles resulting in regulation of nutrient cycling, hydrological flow characteristics, and water retention (*regulating service of the ecosystem*) (Baird and Wilby,

Figure 5.3 The grass strips along the physical structure to deliver 'provisioning' and 'regulating' services to the community.
Photograph by the author.

1999); and (2) for their biomass product such as food, fuel, woods, etc. (*provisioning services of the ecosystem*).

The renovated system resembles a series of 'cross-slope barriers' (Figures 5.2 and 5.3) made of bamboo-matted wooden structure with an average height of 30–40 cm to avoid interference with the cultivation practices. While keeping the principles of ecohydrology, the place-based and use-inspired plantation is to be established along the physical structure in the form of narrow strip. Upon establishment of the biota, the plant strips ecohydrologically overtake the 'regulation' role of the physical structure for long-term functionality.

In its ecohydrological functionality of the renovated system, the anticipated low-cost and high-impact ecological engineering (*the third principle of ecohydrology*) is established to regulate the surface runoff (*the first principle of ecohydrology*). At the establishment stage, the biota benefits from the physical structure by gaining the required vital resources (moisture, nutrients, fertile soils, and organic matters). Once established, the vegetation will, in turn, regulate the runoff flow (*the second principle of ecohydrology*) resulting in dual regulation mechanism which is the basic essence of ecohydrology.

5.4. The technique in practice

The proposed system of surface water flow regulation system has been demonstrated in one the sub-basins of Ethiopian Rift Valley Lakes basin. Figure 5.4 shows how the system operates during a flood event. The sediment accumulation behind the field structure in the aftermath of the flood event is shown by Figure 5.5. Figure 5.6 also shows the favorable farming practices by the local community.

5.5 Theoretical bases for determination of spacing between consecutive flow-regulating structures

Three candidate equations/assumptions are proposed to fix the physical spacing between two consecutive flow-regulating barriers

Figure 5.4 The flow regulation functions of the farm structure during a flood event.
Photograph by the author.

Figure 5.5 Condition of sediment accumulation in the aftermath of an extreme flood event.
Photograph by the author.

(Figure 5.7): spacing based on (1) allowable slope length; (2) allowable shear stress; and (3) seepage-saturation zone coincidence. The upcoming sub-sections describe these different approaches in detail.

5.5.1 Criteria 1: allowable slope length approach

Kibler and Aron (1982) indicated that the maximum sheet flow (portion of the overland flow) length is less than 100 feet, whereas McCuen and Spiess (1995) proposed the limiting criteria for sheet flow length as follows:

$$l = \frac{100\sqrt{S}}{n}$$

where: n = Manning's roughness coefficient
λ = limiting length of flow, ft
S = slope, ft/ft

Figure 5.6 Farming practice by the local community on the accumulated vi-
        tal resources
Photograph by the author.

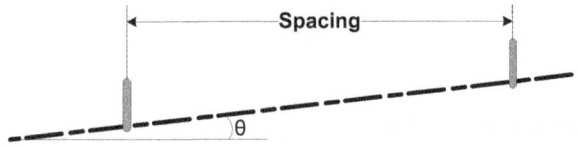

Figure 5.7 Schematics of spacing between two consecutive flow-regulating
        structures.

5.5.2 *Criteria 2: allowable shear stress approach*

The following equations can be used to estimate the critical flow shear
stress for soil detachment (Alberts et al., 1995):
    For cropland surface soils containing $\geq 30\%$ sand:

$$\tau_c = 2.67 + 6.5 \text{ clay} - 5.8 \text{ } vfs$$

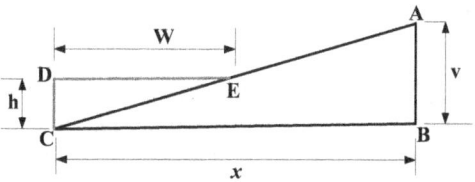

Figure 5.8 Basic diagram for deriving the height of bund.

where clay, vfs are respectively fractions of clay and very fine sand contents in the surface soil and τ_c expressed in Pa. In this equation, vfs is set equal to 0.4 when vfs is <0.4 and clay is set equal to 0.4 if clay is >0.4.

5.5.3 Criteria 3: spacing based on seepage-saturation zone coincidence

This soil moisture-based approach (Figure 5.8) assumes the following criteria WTCER (2011):

i   The phreatic (seepage) zone of the upper bund should coincide with the saturation zone of the immediate lower bund.
ii The bund should be able to check the surface runoff at the point where flow attains an erosive velocity.
iii The bund should satisfy all the requirements of agricultural operations.

5.6 Conclusion and recommendation

Based on the understanding of the potential linkage between eco-hydrology and SLM, this chapter brought a fairly new strategy that potentially narrows down the technical as well as the socioeconomic limitations of some of the actions in the land management sector. For a complete adoption of those strategies, it is recommended to include a comparative cost-benefit analysis with the existing similar techniques.

References

Alberts, E.E., Nearing, M.A., Weltz, M.A., Risse, L.M., Pierson, F.B., Zhang, X.C., Laflen, J.M., & Simanton, J.R. (1995). *Soil component USDA-Water* erosion prediction project: Hillslope profile and watershed model documentation (Rep. 10, chap. 7, pp. 7.1–7.47). West Lafayette, IN: National Soil Erosion Research Lab.

Baird, A.J., & Wilby, R.L. (Eds). (1999). *Ecohydrology. Plants and water in terrestrial and aquatic environments.* London, New York: Routledge.

Belete, M.D. (2021). Review of the underpinning reasons and field demonstrations to incorporate ecohydrologic strategy into landscape restoration in water-limited ecosystems. *Ecohydrology & Hydrobiology, 21*(3), 529–542.

Emmett, W.W. (1970). *The hydraulics of overland flow* (USGS Profession Paper 662-A). Washington, DC: US Government Printing Office.

Falkenmark, M. (2016). Water and human livelihood resilience: a regional-to-global outlook. *International Journal of Water Resources Development, 33*(2), 1–17

FAO (2009). *Country support tool – For scaling-up sustainable land management in sub-Saharan Africa.* Rome: FAO.

Kampf, S.K., & Mirus, B.B. (2013). Subsurface and surface flow leading to channel initiation. In E. Wohl (Ed.), *Treatise on Fluvial Geomorphology.* San Diego, CA: Academic Press.

Kibler, D.F., & Aron, G. (1982). Estimating basin lag and T(c) in small urban watersheds. *EOS, Transactions, American Geophysical Union, 63*(18), abstract # H12-8.

Loewen, A., & Pinheiro, A. (2017). Overland flow generation mechanisms in the Concórdia River basin, in southern Brazil. *Brazilian Journal of Water Resources, 22* (e4).

McCuen, R.H., & Spiess, J.M. (1995). Assessment of kinematic wave time of concentration. *Journal of Hydraulic Engineering, 121* (3), 256–266.

Power, A.G. (2010). Ecosystem services and agriculture: Tradeoffs and synergies. *Philosophical Transactions of The Royal Society B Biological Sciences, 365*(1554), 2959–2971.

Reinders, F.B., Oosthuizen, H., Senzanje, A., Smithers, J.C., Van Der Merwe, R.J., van Der Stoep, I., & van Rensburg, L. (2016). *Development of technical and financial norms and standards for drainage of irrigated lands* (WRC Report No.TT 655/15). Pretoria:Water Research Commission.

Shroder, J., & Ahmadzai, S.J. (2016). *Transboundary water resources in Afghanistan: Climate change and land-use implications* (1st ed.). Amsterdam, The Netherlands: Elsevier Inc.

Sidle, R.C., Jarihani, B., Kaka, S.I., Koci, J., & Al-Shaibani, A. (2019). Hydrogeomorphic processes affecting dryland gully erosion: Implications for modelling. Progress in Physical Geography. *Earth and Environment, 43*(1), 46–64.

Smyth, A.J., & Dumanski, J. (1993). *FESLM: An international framework for evaluating sustainable land management* (A discussion paper. World Soil Resources Report 73). Rome:Food & Agriculture Organization.

Sustainet, E.A. (2010). *Technical manual for farmers and field extension service providers: Conservation agriculture.* Nairobi: Sustainable Agriculture Information Initiative.

Swinton, S., Lupi, F., & Philip, R.G., & Hamilton, S. (2007). Ecosystem services and agriculture: Cultivating agricultural ecosystems for diverse benefits. *Ecological Economics, 64*, 245–252.

Tenge, A.J., de Graaff, J., & Hella, J.P. (2005). Financial efficiency of major soil and water conservation measures in West Usambara highlands Tanzania. *Applied Geography, 25*, 348–366.

Teshome, A., Rolker, D., & de Graaff, J. (2013). Financial viability of soil and water conservation technologies in Northwestern Ethiopian highlands. *Applied Geography, 37*, 139–149.

UN (2021). *Farmlands.* Retrieved from https://www.decadeonrestoration.org/types-ecosystem-restoration/farmlands.

Wood, S., Sebastian, K., & Scherr, S.J. (2001). *Pilot analysis of global ecosystems: Agroecosystems.* Washington, DC: International Food Policy Research Institute and World Resources Institute,

World Bank (1997). *Rural development. From vision to action* (ESSD Studies and Monographs Series 12). Washington, DC: World Bank.

WTCER (Water Technology Centre for Eastern Region) (2011). *Soil and water conservation measures in arable lands.* New Delhi: Natural Resource Management Consultants.

Zalewski, M., & Naiman, R.J. (1985). The regulation of a riverine fish community by a continuum of abiotic-biotic factors. In J.S. Alabaster (Ed.), *Habitat modification and freshwater fisheries.* London: Butterworths Scientific Ltd.

6 Ecohydrologic strategy for restoration of gully networks in a landscape

Mulugeta Dadi Belete

6.1 Introduction

Gully erosion is defined as erosion in channels that are too deep to cross with farm equipment (Hutchinson and Pritchard, 1976) or they are erosional features larger than rills (which are <0.3m deep which can be plowed out or easily crossed) but smaller than streams, creeks, arroyos, or river channels (Wells, 2004). Gullies are formed by a concentrated flow of water removing topsoil as well as parent material (SCS, 1966). Gully erosion is one of the most active geomorphic processes (Medvedeva, 2018) and a major environmental problem, posing significant threats to sustainable development (Frankl et al., 2020). It is an important signature of landscape degradation and represents the worst stage of all types of soil erosion. Gully erosion is also a highly visible form of erosion (Abdulfatai et al., 2014) that generates between 10% and 95% of total sediment mass at watershed scale whereas gully channels usually occupy less than 5% of the total watershed (Poesen et al., 2003). In addition, they damage agricultural fields/infrastructure, alter transportation corridors, and degrade surface water quality (Takken et al., 2008). Gullies occur worldwide across a wide range of climatic, geomorphological, and pedological conditions (Billi and Dramis, 2003) and they produce large volumes of sediment that are eventually transported downstream affecting the water quality, reservoir capacity, and floodplains (Wells et al., 2010). If left untreated, they can eventually turn a productive landscape into degraded ecosystem that will cost much for its restoration.

Significance of gullies as a potential focus for landscape enhancement was highlighted as early as 1972 (McLeary, 1972). The benefits of gully restoration are numerous and wide ranging including improved environmental, aesthetic, scenic, and cultural values (Clarkson and Mcqueen, 2004). However, the ultimate goal of gully management is to stabilize the gully and rehabilitate vegetation (Liu et al., 2019).

DOI: 10.4324/9781003309130-6

According to USDA (2007), gully treatment can be done in three stages namely: stabilization, rehabilitation, and restoration. The primary treatment is considered as *'stabilization'* which targets halting the expansion of erosion and gully networks, reducing sediment yield, and improving water quality. On the other hand, *'rehabilitation'* targets not only to stop the erosion expansion but also improves other resources such as timber, recreation, and wildlife. Finally, *'restoration'* comprises a more comprehensive effort to return the affected land to an acceptable condition for hydrologic, soil productivity, and biologic responses. Most of these targets involve interactions among water-soil-plant for their effectiveness. Among treatment goals of gullies, the 'restoration' concept is implicitly tied to the ecohydrologic functionality of the gully ecosystem as it relies on the hydrology-soil-biota interplay. However, explicit consideration of this interaction as a management tool is very limited (both in theory and practice) in the field of erosion control in general and gully restoration in particular.

Inspired by the above realities and with recognition of the important water-soil-plant interaction governed by the (terrestrial) ecohydrologic principles, this chapter theoretically proposes and practically demonstrates various ways of achieving ecohydrological functionality of a given gully ecosystem by enhancing the absorbing capacity of the ecosystem through the application of ecohydrology-based landscape restoration (EcoLaR) approach, as described in the upcoming sections. The techniques operate according to the three principles of ecohydrology (Zalewski et al., 1997) as follows:

1 In line with the first principle of ecohydrology (*hydrological framework*), the hydrology (surface water run-off) in the gully system is going to be regulated as a template for functional integration of hydrological and biological processes.

2 According to the second principle (*ecological target*), the techniques target enhancement of the carrying capacity and ecosystem services of the gully system through place-based and need-driven plantation (biota establishment). This biota establishment in the gully system is governed by the degree of hydrologic regulation that ensures water availability as well as regulation of nutrient cycle.

3 The techniques integrate the above hydrological framework and ecological targets to restore the gully ecosystem as a new tool for restoration of gully ecosystem (*ecological engineering method*).

The approach proposes different interventions for different sections of the gully system: gully head, gully bed, and gully banks. The upcoming sections illustrate the proposed techniques from both theoretical and practical perspectives.

6.2 Bamboo-matted plunge pools as energy dissipater of water jets at sharp gully heads

Basically, the run-off that enters the gully system is in motion (has kinetic energy) and creates waterfalls (has potential energy) that cause further erosion on the gully head due to the jet impact of overland flow from the upstream (Dey et al., 2007). These energy forms can cause further erosion leading to further instability of the gully system. Hence, the first step to control gully system in general and gully heads (Figure 6.1) in particular is to dissipate the flow energy, which is a major issue in hydraulic engineering (Xu etal., 2015). In order to take advantage of dam engineering, which usually entails

Figure 6.1 An example of a gully head with very vertical head in the middle of farmlands.
Photograph by the author.

energy dissipation mechanisms, this chapter proposes the 'plunge-pool' system (Figure 6.2) to play the anticipated ecohydrologic role. The proposed plunge-pool system is modified in such a way that the energy dissipation (hydrological regulation) as well as the vegetation growth (biota response) operates ecohydrologically. That means, the proposed system is equivalent to the extrapolation of the concept of ABRC – Abiotic Biotic Regulatory Concept (Zalewski and Naiman, 1985) in which the (hydrologic) regulation in the form of 'energy dissipation' represents the abiotic dimension while establishment of vegetation represents the biotic response. In line with this concept, the hydrologic regulation is considered as a prerequisite for the establishment of vegetation (the biota) in the gully system. Once the vegetation in the gully system is established, the dual regulation will take place. Using this interaction as a background tool for the management of the gully system qualifies the principles of ecohydrology.

In the EcoLaR approach, the bamboo-matted plunge pool (Figure 6.2) is proposed in an attempt to provide a means to absorb or dissipate the energy from the inflow discharge jet and protect the gully head from erosion and undermining.

It is the plunge pool that absorbs most of the energy in the water jets so that when the water flow reaches the gully bed, it has lost most of its scouring ability (Puertas and Dolz, 2005). This phenomenon is the

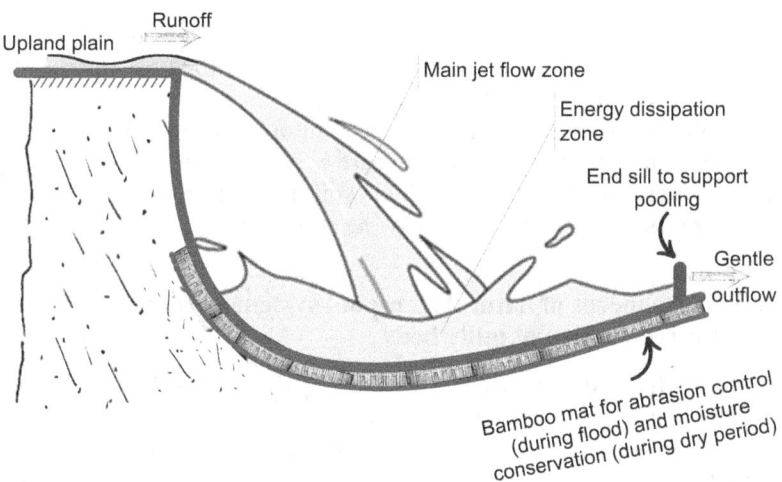

Figure 6.2 Schematics of a bamboo-matted plunge pool at a gully head.

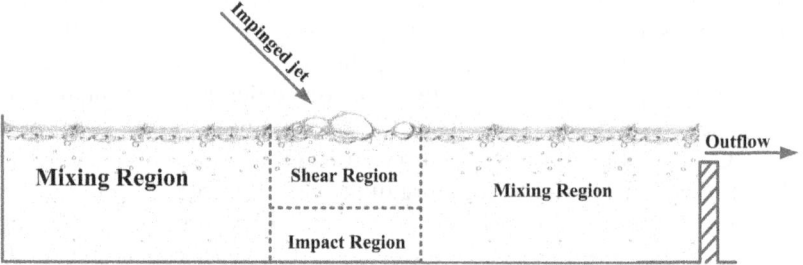

Figure 6.3 The general flow characteristics of a plunge pool in different regions of energy dissipation in the anticipated plunge-pool system (Modified after Puertas and Dolz, 2005).

result of energy transferring in which the water jet impinging into the plunge pool loses energy when the time-averaged energy of the jet is turned into turbulent energy. This energy dissipation system has three regions (Figure 6.3): shear region (where the jet hits the plunge pool); impact regions (where the energy is dissipated by impact on the pool floor); and mixing region (where the energy is dissipated by turbulent mixing before the water leaves the plunge pool) (Xu et al., 2002).

In practice (Figure 6.4), the behavior of the jet in the air, in the mass of water and pressures and forces on the plunge-pool floor have to be taken into account (Khatsuria, 2004) for proper design of the system. Xu et al. (2002) points out that the pressure on the plunge-pool floor is the most important parameter when designing plunge pools in which the maximum pressure as well as the gradient of the pressure distribution on the pool floor should be restricted.

Ecohydrologically, the proposed bamboo-matted plunge-pool system facilitates biota responses (Figure 6.5) by increasing the soil moisture content of the ecosystem. This soil moisture-plant interaction is the core issue of ecohydrology (Wang et al., 2019).

6.3 The concept of natural step-pool system as a strategy for restoration of gully beds

Gully beds are usually treated with check dams which are designed to enhance sediment deposition reducing the bed gradient and flow velocity in order to check soil erosion within the system (Conesa-García and Lenzi, 2010). However, these structures should be as compatible as possible with the channel's natural tendency to reach a stable

Figure 6.4 An example of vertically sharp edged head treated with bamboo-matted plunge.
Photograph by the author.

Figure 6.5 Vegetative feedbacks due to moisture enhancement and protection
against water jet energy by the structure.
Photograph by the author.

configuration over a long period of time (Lenzi, 2002). In order to assess the degree of compatibility of those interventions with nature, it
is of paramount importance to observe how natural streams operate
to stabilize themselves. At this juncture, the concept of step-pool system appears to be the logical phenomenon to deal with. Step-pools are
channel forms composed of alternating channel-spanning ribs (steps)
and pools with tumbling flow (Peterson and Mohanty, 1960) that os-
cillates between subcritical in the pool and supercritical over the step
(Grant et al., 1990).

6.3.1 How the natural step-pool system operates to stabilize
stream beds?

Step-pool configuration of stream beds represents an interesting case
of spontaneous, self-organized system of high stability (Chin and Phillips, 2007) that is formed by infrequent high flows that pick up large
rocks and logs and carry them down rivers. Once the water level drops,

the debris is deposited in the stream (Chin, 1989). When these objects are large relative to the size of the channel, pools form behind them, and a step-pool-step pattern develops (Hayward, 1980). This morphology is known to be a configuration that effectively dissipates energy (Chin, 2003) and tends to maximize flow resistance, leading to minimum velocity and shear stress, which is the final cause of its stability (Abrahams et al., 1995).

One of the main roles of this self-organizing system of natural step-pools in stream channels is to stabilize stream beds and reduce erosion during heavy flooding events because the steps increase the flow resistance, consume the flow energy, and protect the streambed from erosion (Church and Zimmermann, 2007). This capacity of step-pool configuration to modulate the sediment fluxes is used for ecological and management issues (Pellegrini et al., 2021). The energy dissipation mechanism of the step-pool system takes place vertically rather than horizontally, like what is found in streams and rivers closer to sea level. As the water cascades over the steps, its horizontal kinetic energy is turned into vertical energy and absorbed as turbulence in the pool below (Clarke, 1988).

In the context of gully restoration, the analogy between step-pool systems and check dam interventions has been recently recognized, to the point that check dams have been considered as the anthropogenic equivalent to step-pool sequences in steep mountain streams. In spite of the suggested similarities, the maximum flow resistance in step-pool series occurred with significantly shorter spacing than that recommended for gully control using check dams (Heede, 1976; Morgan, 2005). In some cases, their morphological features have inspired the design criteria for artificial check dam sequences in high-gradient streams stabilization (Lenzi, 2002; Wang and Yu, 2007; Conesa-García and Lenzi, 2010).

Inspired by the functionality of the natural step-pool system on stream beds, the EcoLaR approach attempts to shift from the conventional check dam system of gully bed control into this ecohydrologic strategy of restoration of gully beds (Figure 6.6). The anticipated system regulates the hydrology in the gully (*EH principle 1*) that in turn triggers biota responses (*EH principle 2*) through which the hydrology-biota regulation (*EH principle 3*) will take place. The energy-reducing mechanism is primarily the head loss from penetrating the pool system twice – first to get into the bank zone and again to get out. In this system, sediment is preferentially stored in pools (Marion and Weirich, 1999) as the step-pools are effective energy dissipaters (Chin, 2003) and provide potential habitat (Scheuerlein, 1999). In terms of the ecosystem services, this intervention provides: run-off flow regulating functions; sediment trapping functions; moisture conserving functions; and provision of suitable site for plantation along the gully network (Figures 6.6-6.9).

According to Whittaker and Jaeggi (1982), the role of the step-pool configuration as energy dissipaters can be impaired when pools get filled with sediment (Figure 6.8). In this case, additional layers on top of the already filled pool may be required.

Upon successful regulation of the run-off (EH principle 1) in the gully system, the second EH principle will guide what to plant along the gully system. The EcoLaR approach recommends to plant place-based and need-driven vegetation (Figure 6.9) to benefit from the regulated flow as well as the improved condition of soil moisture. Over time, these plants in turn regulate the hydrology and fulfill the third principle of EH.

6.3.2 Design criteria for step-pool system at gully bed

Spacing and height are the two design parameters for the proposed step-pool system. In the natural set-up, the length of a regular step-pool system, or the distance between two steps or two pools is inversely proportional to the average slope of the stream and increases with the average discharge (Whittaker, 1987). Abrahams et al. (1995) also noted that the ratio of the step height to the length is proportional to the average slope. Curran and Wohl (2003) and Chartrand and Whiting (2000) scaled the spacing between two consecutive steps or pools to be in the range of one to four channel widths. Similarly, Grant et al.

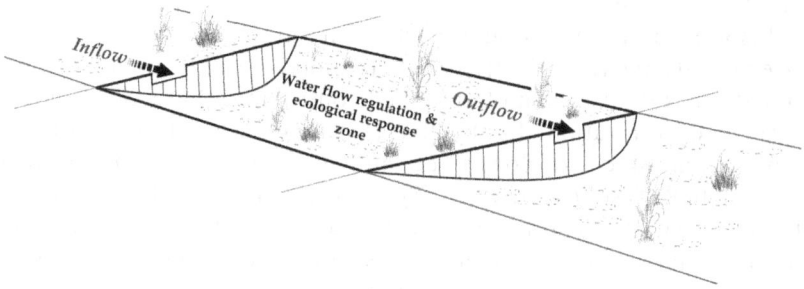

Figure 6.6 Schematics of the stilling pool as hydrologic regulator at gully beds.

Figure 6.7 Step-pool system along the gully bed.
Photograph by the author.

Figure 6.8 The step-pool system during a flood event.
Photograph by the author.

Figure 6.9 The step-pool system to facilitate in-stream vegetation.
Photograph by the author.

(1990) recommended the spacing to be in the magnitude of 2–4 folds of the channel widths.

In general, we can adopt either of the mathematical relationships given by Abrahams et al. (1995) or Chin et al. (2008), whichever is smaller as presented below:

$H/L = 1.5S$ (Abrahams et al., 1995); where H is the step height (m), L is the distance between the steps (m), and S is the gradient of streambed.

Or $L = 7.39 \ln (d_{step}/S)-5.52$ (Chin et al., 2008);

where L is the step spacing (in meters) and d_{step} is step height (in meters).

In order to release surplus water from the pools, we need to provide a spillway using the well-known weir equation:

$$Q = CLD^{3/2}$$

where C is the coefficient taken as 3.0, L is the length of spillway (m), D is the depth of spillway (m), and Q is the maximum discharge of gully catchment at the proposed site (m^3/s).

6.4 Strategy to rehabilitate sharp curves of gully beds

Sharp curves of meandering gullies are critical points of attack by which the gully banks are expanded and/or collapsed. The technical objectives of rehabilitating these critical locations may comprise:

- Moving the area of deepest, fastest flow away from the river banks
- Providing ecological habitats and enhance biodiversity along the gully continuum
- Preventing bank erosion and controlling flow direction and strength
- Creating an erosion free zone for biota establishment
- Enhancing gully structure
- Concentrating the flow into the gully axis
- Restoring geomorphology of the gully

In order to achieve the above design objectives, the EcoLaR approach proposes the use of spurs (sometimes named as spur dykes, groynes, dykes, or transverse dykes) (Figure 6.10).

6.4.1 How spurs operate hydraulically

As shown in Figure 6.11, the protection mechanism relies on the inability of water to flow to a sharp channel bank that gives an opportunity for the gully bank to get protected for certain distance upstream and downstream of the groynes. The action of eddies reduces from the

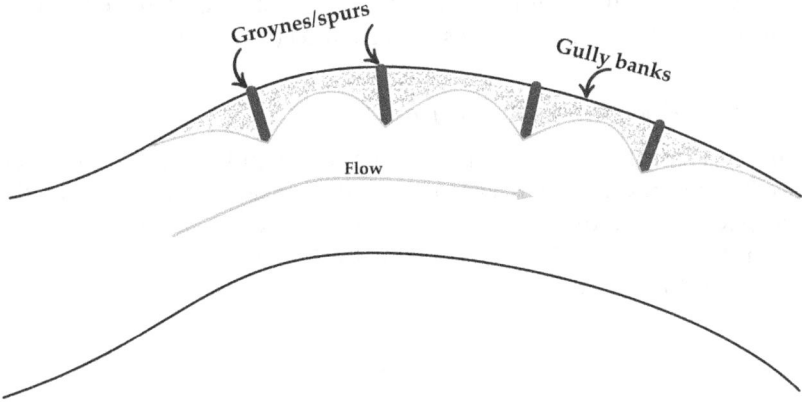

Figure 6.10 Schematics of spurs at gully bend.

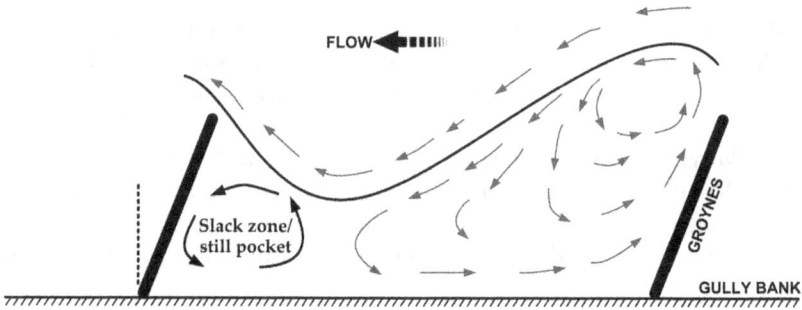

Figure 6.11 Schematics of flow pattern between two sequential repelling type groynes.

head toward the bank. In adopting the concept of spur for gully re-habilitation, it is preferred to use the 'permeable' type for helpfulness in causing quick siltation due to damping of velocity. They are useful when concentration of suspended sediment load is heavy while allowing water to pass through.

6.4.2 Design criteria for spur length

The three most important design parameters of spurs are: length, spacing, and angle. When choosing the length of a spur, it is important to consider the safety of the opposite bank. If a spur is too long, it may guide the river current during a flash flood to the opposite bank which will cause damage; if it is too short, it may cause erosion of the near bank. Normally the effective length of spur shouldn't exceed 1/5th of the width of flow in case of a single channel (CWC, 2012).

6.4.3 Design criteria for spur spacing

It is evident that spurs can protect a reach of gully bank to the extent of 2–2.5 times their own length protruding inside the gully. Generally, a spacing of 2.5 times the length of the groyne is generally adopted at convex banks, while spacing equal to the length of the groyne is mostly adopted for concave banks.

6.4.4 Design criteria for spur angles

Generally, repelling type or deflecting spurs are provided for anti-erosion measures. Repelling type spurs may be kept at an angle of 50–100 against the direction of flow (CWC, 2012).

6.4.5 Vegetating the slack zone for ecohydrologic feedback

The spurs create a zone of slack flow (retention areas) which encourages silting up in the region of the spur to create a natural bank. It was evident that nutrient transformation and retention that take place in these zones present a green service which has a socioeconomic importance (Gren et al., 1995). They generally protect the riparian environment and often improve the pool habitat and physical diversity. Functioning as storage zones, they are also key elements during lower water tables and can provide substantial support to the food web (Hein et al., 2005).

6.5 Application of 'regime theory' to narrow down gully widths and provision of plantation sites for additional ecosystem services

There are cases where a given reach along a gully network becomes unusually wide as shown in Figure 6.12.

In the EcoLaR approach, this unusually wide reach of the gully is considered as an opportunity for in-stream plantation (Figure 6.13).

Figure 6.12 A wide gully reach to be narrowed down by guide banks.
Photograph by the author.

Figure 6.13 Local tradition of in-stream plantation @ Lake Hawassa Watershed. Photograph by the author.

As shown in Figure 6.14, the reach is subjected to narrowing down by guide banks.

Following Lacey's Regime Theory, the linear water way is proposed to be fixed using the formula that is claimed to be Lacey's formula (Lacey, 1929, 1933):

$$L = 4.83 \ Q$$

in which L is the allowable gully width in the given reach, the regime perimeter in m and Q the design flood discharge of the river in m^3/s.

6.6 Ecohydrological manipulation of gully banks

With respect to the principles of ecological engineering, the EcoLaR approach is determined to restore the landscape with minimum earth

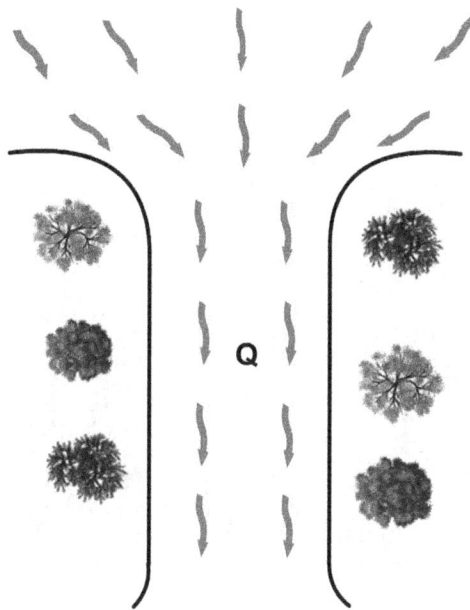

Figure 6.14 Schematics of a guide bank to narrow down gully width to create plantation site.

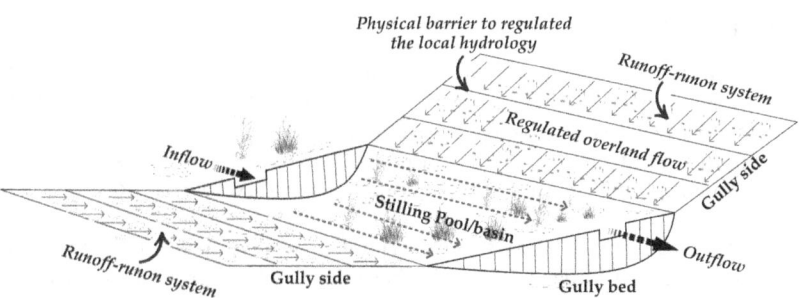

Figure 6.15 Schematics of gully side rehabilitation system of EcoLaR approach.

work. A similar concept is applied to rehabilitate gully banks as the approach conceives the ecohydrological opportunities of the gully banks. By applying the basic structural unit of the green infrastructure (Figure 3.1), the gully banks are going to be turned into plantation site (Figures 6.15 and 6.16).

Figure 6.16 Gully bank rehabilitation technique with minimum earth work.
Photograph by the author.

References

Abdulfatai, I.A.O., Akande, W.G., Momoh, L.O., & Ibrahim, K.O. (2014). Review of gully erosion in Nigeria: causes, impacts and possible solutions. *Journal of Geosciences and Geomatics, 2*(3), 125–129.

Abrahams, A.D., Li, G., & Atkinson, J.F. (1995). Step-pool streams: adjustment to maximum flow resistance. *Water Resources Research, 31*, 2593–2602.

Billi, P., & Dramis, F. (2003). Geomorphological investigation on gully erosion in the Rift Valley and the northern highlands of Ethiopia. *Catena, 50*, 353–368.

Chartrand, S.M., & Whiting, P.J. (2000). Alluvial architecture in headwater streams with special emphasis on step-pool topography. *Earth Surface Processes and Landforms, 25*, 583–600.

Chin, A. (1989). Step pools in stream channels. *Progress in Physical Geography: Earth and Environment, 13*(3), 391–407.

Chin, A. (2003). The geomorphic significance of step–pools in mountain streams. *Geomorphology, 55*, 125–137.

Chin, A., & Phillips, J. (2007). The self-organization of step-pools in mountain streams. *Geomorphology, 83*, 346–358.

Chin, A., Anderson, S., Collison, A., Ellis-Sugai, B.J., Haltiner, J.P., Hogervorst, J.B., & Wohl, E. (2008). Linking theory and practice for restoration of step-pool streams. *Environmental Management, 43*(4), 645–661.

Chow, V.T. (1959). *Open channel hydraulics.* New York: McGraw.

Church, M., & Zimmermann, A. (2007). Form and stability of step-pool channels: Research progress. *Water Resources Research, 43*(3), W03415.

Clarke, A.J. (1988). Inertial wind path and sea surface temperature patterns near the Gulf of Tehuantepec and the Gulf of Papagayo. *Journal of Geophysical Research-Oceans, 93*, 15491–15501.

Clarkson, B.D., & McQueen, J.C. (2004). *Ecological restoration in Hamilton City, North Island, New Zealand.* In proceedings of the 16th international conference, society for ecological restoration. Victoria, Canada.

Conesa-García, C., & Lenzi, M.A. (2010). *Check dams, morphological adjustments and erosion control in torrential streams.* Hauppauge, NY: Nova Science Publishers.

Curran, J.H., & Wohl, E. (2003). Large woody debris and flow resistance instep-pool channels. *Geomorphology, 51*, 141–157.

CWC (Central Water Commission) (2012). *Handbook for flood protection, anti-erosion and river training works.* India: Central Water Commission.

Dey, A.K., Tsujimoto, T., & Kitamura, T. (2007). Experimental investigations on different modes of headcut migration. *Journal of Hydraulic Research, 45*, 333–346.

Frankl, A., Nyssen, J., Vanmaercke, M., & Poesen, J. (2020). Gully prevention and control: techniques, failures and effectiveness. *Earth Surface Processes and Landforms, 46*(1), 220–238.

Grant, G.E., Swanson, F.J., & Wolman, M.G. (1990). Pattern and origin of stepped-bed morphology in high-gradient streams, Western Cascades, Oregon. *Geological Society of America Bulletin, 102*, 340–352.

Gren, I.M., Groth, K.H., & Sylven, M. (1995). Economic values of Danube floodplains. *Journal of Environmental Management, 45*, 333–345.

Hayward, J.A. (1980). *Hydrology and stream sediments in a mountain catchment* (Canterbury, New Zealand: Tussock Grasslands and Mountain Institute Special Publication, 17).

Heede, B.H. (1976). *Gully development and control: the status of our knowledge* (Research paper RM 169). Fort Collins, CO: Forest Service, USDA.

Hein, T., Reckendorfer, W., Thorp, J. H., & Schiemer, F. (2005). The role of slack water areas for biogeochemical processes in rehabilitated river corridors: Examples from the Danube. *Large Rivers, 15*(1–4), 425–442.

Hutchinson, D.E., & Pritchard, H.W. (1976). Resource conservation glossary. *Journal of Soil and Water Conservation, 31*, 1–63.

Khatsuria, R.M. (2004). *Hydraulics of spillways and energy dissipators.* Boca Raton, FL: CRC press.

Lacey, G. (1929). Stable channels in alluvium. Proceedings Institution of Civil Engineers. *Proceedings Paper, 4736* (229).

Lacey, G. (1933). Uniform flow in alluvial rivers and canals. *Minutes of the proceedings of the institution of civil engineers, 237*, 421–453.

Lacey, G. (1958). Flow in channels with sandy mobile beds. *Minutes of the proceedings of the institution of civil engineers, 9*, 145–164.

Lenzi, M.A. (2002). Stream bed stabilization using boulder check dams that mimic step-pool morphology features in Northern Italy. *Geomorphology, 45*, 243–260.

Liu, X., Li, H., Zhang, S., Cruse, R.M., & Zhang, X. (2019). Gully erosion control practices in Northeast China: A review. *Sustainability, 11*(18), 5065.

Marion, D.A., & Weirich, F. (1999). Fine-grained bed patch response to near-bankfull flows in a step-pool channel. In D.S. Olsen, J.P. Potyondy (Eds.), *Wildlife hydrology* (American water resources association technical publication series No. 99-3). Bethesda, MD: American Water Resources Association.

McLeary, W.H. (1972). *A study of the gully systems of the Waikato Basin with particular reference to those in and surrounding the City of Hamilton* (Masters thesis, University of Canterbury-Lincoln College).

Medvedeva, R.A. (2018). Trends of the gully erosion development in the territory of the Republic of Tatarstan. *IOP Conference Series Earth and Environmental Science, 107*(1), 012016.

Morgan, R.P.C. (2005). *Soil erosion and conservation* (3rd ed.). Cornwall: Black-well Publishing.

USDA (2007). Gullies and their control, technical supplement 14P. National Engineering Handbook Part 654 (210–VI–NEH). USA.

Pellegrini, G., Rainato, R., Martini, L., & Picco, L. (2021). The morphological evolution of a step-pool stream after an exceptional flood and subsequent ordinary flow conditions. *Water, 13*, 3630.

Peterson, D.F., & Mohanty, P.K. (1960). Flume studies of flow in steep, rough channels. *American Society of Civil Engineers, 86*, 55–76.

Poesen, J., Nachtergaele, J., Verstraeten, G., & Valentin, C. (2003). Gully erosion and environmental change: Importance and research needs. *Catena, 50*(2–4), 91–133.

Puertas, J., & Dolz, J. (2005). Plunge pool pressures due to a falling rectangular jet. *Journal of Hydraulic Engineering, 131*(5), 404–407.

Scheuerlein, H. (1999). Morphological dynamics of step-pool systems in mountain streams and their importance for riparian ecosystems. In A.W. Jayawardena, J.H. Lee, Z.Y. Wang (Eds.), *River sedimentation: Theory and applications*. Rotter-dam: A.A. Balkema.

SCS (1966). *Procedure for determining rates of 633 land damage, land depreciation, and volume of sediment produced by gully erosion* (Technical Release No. 32. US 634 GPO 1990-261-419:20727/SCS). Washington, DC: US Government Printing Office.

Singh, V.P. (2003). Theories of hydraulic geometry. *International Journal of Sediment Research, 18*(3), 196–218.

Takken, I., Croke, J., & Lane, P. (2008). Thresholds for channel initiation at road drain outlets. *Catena, 75*, 257–267.

Wang, C., Fu, B., Zhang, L., & Xu, Z. (2019). Soil moisture–plant interactions: An ecohydrological review. *Journal of Soils and Sediments, 19*, 1–9.

Wells, N.A. (2004). Gully. In A. Goudie (Ed.), *Encyclopedia of geomorphology.* IAG, USA and Canada: Routledge.

Wells, R.R., Bennett, S.J., & Alonso, C.V. (2010). Modulation of headcut soil erosion in rills due to upstream sediment loads. *Water Resources Research, 46*, 2–16.

Whittaker, J.G. (1987). Sediment transport in step-pool streams. In C.R. Thorne, J.C. Bathurst, & R.D. Hey (Eds.), *Sediment transport in gravel-bed rivers.* Chichester: Wiley.

Whittaker, J.G., & Jaeggi, M.N.R. (1982). Origin of step-pool systems in mountain streams. *American Society of Civil Engineers, 108*, 758–773.

Xu, W., Liao, H., Yang, Y., & Wu, C. (2002). Turbulent flow and energy dissipation in plunge pool of high arch dam. *Journal of Hydraulic Research, 40*(4), 471–476.

Xu, W.L., Luo, S.J., Zheng, Q.W., & Luo, J. (2015). Experimental study on pressure and aeration characteristics in stepped chute flows. *Science China Technological Sciences, 58*, 720–726.

Zalewski, M., & Naiman, R.J. (1985). The regulation of a riverine fish community by a continuum of abiotic-biotic factors. In J.S. Alabaster (Ed.), *Habitat modification and freshwater fisheries.* London: Butterworths Scientific Ltd.

Zalewski, M., Janauer, G.A., & Jolankai, G. (1997). *Ecohydrology. A new paradigm for the sustainable use of aquatic resources* (Technical Document in Hydrology No. 7; IHP – V Projects 2.3/2.4). Paris: UNESCO IHP.

7 Characterization of vegetated riparian buffer zones as the last line of defense for protecting water bodies from degradation

Meta-analysis

Mulugeta Dadi Belete

7.1 Introduction

Historically, humankind has tended to settle near streams because of the role of rivers as transportation corridors and the fertility of riparian areas (Di Baldassarre et al., 2013). The term 'riparian' is derived from the Latin word 'riparius', meaning 'of or belonging to the bank of a river' (Naiman and Decamps, 1997) or simply 'stream bank' (Coats, 1999). From an ecohydrological perspective, a riparian zone is the area adjacent to a stream that is subject to direct influence of the water in the stream (Coats, 1999). These ecotones are located at the interface of aquatic (*streams, lakes, reservoirs, rivers, and wetlands*) and terrestrial ecosystems (*usually human-disturbed lands*) that link and influence the ecological functioning of both ecosystems (Ma, 2016). These zones perform ecological functions that are distinct from other components of the landscape (Jones and Stokes, 2005).

Riparian buffers have been generally seen as a minor component of the entire landscape system (Wilcox et al., 2017) despite their supporting ecosystem services as habitats of fauna and flora and critical providers of ecosystem services to the watershed inhabitants (Soykan and Sabo, 2009). They are one of the most commonly used solutions for protecting water quality in rivers, streams, lakes, and reservoirs (Mander et al., 2017). These zones have distinct soil, hydrology, biophysical conditions, and ecological processes (Naiman et al., 2005) as they encompass hydrogeomorphic, vegetational, and food-web attributes (Gregory et al., 1991). They are also considered as effective and sustainable means of buffering aquatic ecosystems against nutrient

DOI: 10.4324/9781003309130-7

stressors (Phillips, 1989) termed as 'the last line of defense' by practitioners for protecting water bodies from degradation (Fortier et al.,
2010). The importance of riparian zones far exceeds their proportion
of land cover (Leroux and Loreau, 2008), and they are vital and fun-
damentally important (Lind et al., 2019) elements of watersheds. Thus,
protecting the riparian buffers is the best and most economical way
to bar non-point source pollution from surface waters (Merrill, 2016)
with a large effect on water quality relative to their size (Quinn, 2003).
Despite the above attributes, the scientific evidences for their alleged
benefits are often lacking and scantily available to policymakers and
practitioners.

Management of the riparian buffers usually entails four general objectives: erosion control; water quality enhancement; aquatic habitat;
and terrestrial habitat improvement (Hawes and Smith, 2005; Lind
et al., 2019) in a cost-effective way (Palone and Todd, 1998). If designed
correctly, buffer zones can also provide important steps toward more
functional land management by: providing flood protection and habitat for animals and plants; increasing ecological connectivity; and creating recreational areas (Mankin et al., 2007); stabilizing soils; slowing
the velocity of runoff to trap sediment and enhancing infiltration of
runoff water; and increasing the uptake and transformation of dissolved nutrients in subsurface waters (Schultz et al., 2009).

Ecohydrologically, they are increasingly portrayed as an important environmental management tool (Wilcock et al., 2009), and have
important implications for water quality and biodiversity (Renouf
and Harding, 2015) while also enhancing the physical, chemical, and
biological integrity of the terrestrial and aquatic ecosystems. A good
example of terrestrial ecohydrologic solutions is creation of these
land-water transition zones (ecotones) for reduction of pollution
fluxes from land to waters from diffused sources (Zalewski, 2014).

As an integral part of the ecohydrology-based landscape restoration
(EcoLaR) approach, this chapter is aimed at providing scientific information on the use of riparian buffer zone as a tool for environmental
management. A review of the underlying ecohydrological processes
that enable the vegetated riparian buffer zone to serve as management
tool or 'the last line of defense' in protecting water resources from
degradation will be presented. The chapter also aims at conducting
meta-analysis to characterize vegetated riparian buffers in terms of
their optimum zonation patterns; widths of the vegetated buffer; and
their efficacy in removing sediment, nitrogen, nitrate, phosphorus,
pollutants, and suspended solids.

7.2 Methods

The single and/or combination search terms including riparian buffer, vegetated buffers, zone, strip, width, sediment, nitrogen, nitrate, phosphorus, pollutants, suspended solids, efficacy, removal efficiency, etc. were used within the database search engines listed here to locate riparian buffer zone literature: Academic Search Premier; Agricola; Biosis Citations Index; Directory of Open Access Journals; PubMed; Scopus; and Web of Science Core Collections.

7.3 Findings

7.3.1 The underlying ecohydrologic processes in the riparian buffers that are serving as management tools

By virtue of their location between uplands and receiving waters, riparian buffer zones are recognized as 'green' infrastructure for performing vital ecohydrologic processes (Figure 7.1) that are closely connected to many water-related ecosystem services (Sun et al., 2015). These services include: the 'interception' of surface runoff, wastewater,

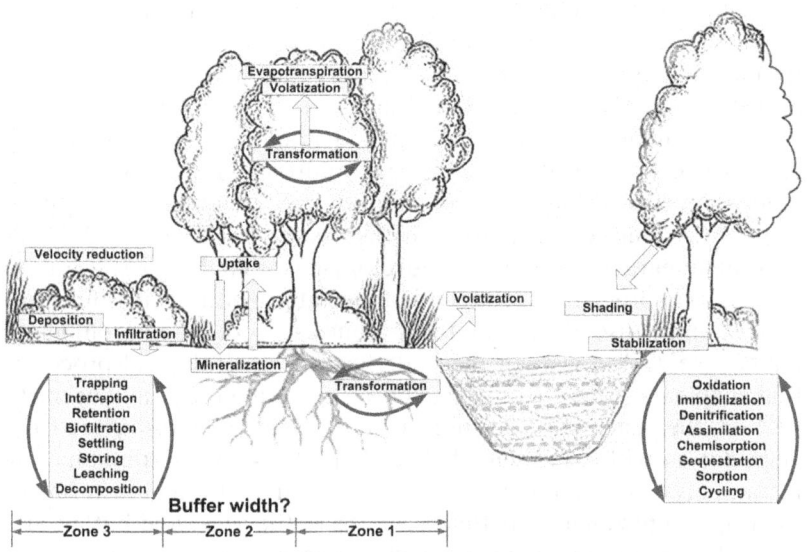

Figure 7.1 Schematics of ecohydrologic processes through which riparian buffers benefits the ecosystem.

subsurface flow, and deeper groundwater flows from upland sources (Welsch, 1991); filtering, retention, and storage of fresh water (Fischer and Fischenich, 2000); storing and recycling of nutrients such as N and P and organic matter (Barling and Moore, 1994); removing or breaking down of xenic nutrients and compounds (Castelle et al., 1994); chemical uptake by plants (Vidon et al., 2010); thereby enhancing quality of receiving water bodies by significantly reducing the amount of removal of nutrients such as nitrogen, phosphorous, and sediment from surface and subsurface flow (Hill, 1996). Riparian buffers also provide shade, shelter, and food for fish and other aquatic organisms; wildlife habitat; and economic products by decreasing soil erosion (Castelle et al., 1994).

Shading is also important to control and favor the development of 'clean-water' invertebrate communities (Fenemor and Samarasinghe, 2020). In addition, riparian buffers provide suitable reproduction habitats for aquatic organisms and amphibians (de Groot, 1992) and clean breathable air by regulating the 'bio- geochemical cycles' (Wilson et al., 2005). In a landscape context, riparian buffers are analogous to kidneys because they filter surface and subsurface inputs and reduce sediment and contaminant transport (Pinay et al., 2018). Nutrients are absorbed into the buffer sediment, taken up by plant biomass, and immobilized by microorganisms through denitrification (Hruby, 2013).

7.3.2 Relationship between widths of buffer strips and their efficacy for different purposes

Conventionally, vegetated buffer strips (VBS) are established as a sole management practice in the riparian buffers which comprises trees, herbs, and grasses (Uusi-Kämppä et al., 2012) where its effectiveness depends on a variety of factors (Burdon et al., 2020). However, buffer width is the most important controllable variable in determining the effectiveness of buffers in reducing pollutants and protecting stream health (Lam et al., 2011). Therefore, identification of optimal riparian buffer width is an important policy issue in watershed conservation (Ekness and Randhir, 2007) and riparian policies also typically prescribe a minimum width for protection (Luke et al., 2018).

Trapping efficiency of buffers is generally improved when the width of the buffer is increased (Barfield et al., 1998) alongside increased vegetation quality (Luke et al., 2018). However, a one-size-fits-all approach is unlikely to be optimal in any particular situation (Everett, 2003) and insufficient (Luke et al., 2018). Moreover, recent approaches tend to disregard fixed-width buffers as they can be grossly inaccurate due to the poor and inconsistent relationship between riparian width and its ecological functionality (Abood & Maclean, 2011). The variable buffer width is collectively supported by previous studies for it greatly depends on circumstances (Hansen et al., 2015) such as the type of water resource we are trying to protect (Hawes and Smith, 2005) and the kind of anticipated ecosystem function (Lind et al., 2019). However, Castelle et al. (1994) reported that a buffer width that is less than 10m is generally ineffective.

There has also been considerable debate over minimum riparian buffer width for protection of stream ecosystems (Sweeney and Newbold, 2014) and the fact that effective buffer width varies by function, so context-specific recommendations are needed (Luke et al., 2018). The following sections present the relationship between buffer widths with their efficacy for sediment retention and nitrogen, nitrate, and phosphorus removal.

7.3.3 *Efficacy of riparian buffers on sediment retention*

Ecohydrologically, the riparian zone can slow (and spread) the incoming overland flow allowing suspended sediments to settle out or be adsorbed into the soil. This regulation gives opportunity for riparian vegetation to use some of the retained nitrogen and phosphorus to grow (Fenemor and Samarasinghe, 2020).

In line with the beneficial role of riparian vegetation in terms of sediment retention described by Daniels and Gilliam (1996), Dosskey et al. (2010), McKergow et al. (2003) and many others, it is observable that the average sediment removal efficiency of vegetated buffers is to the magnitude of 78% with average buffer width of 14 m [*N = 114 sites*] (Figure 7.2). Ecohydrologically, such sediment flux reduction effect predominantly results from flow retardation, filtering, interception, settling, trapping, storage, and deposition processes taking place in the buffer zone.

7.3.4 *Efficacy of vegetated riparian buffers on nitrogen removal*

Through the dominant ecohydrological mechanisms of microbial denitrification (Boyer et al., 2006), nitrogen cycling, and plant sequestration, the nitrogen buffering capacity of vegetated buffers is to the average magnitude of 73% with average buffer width of 15 m (*N = 66 sites*) (Figure 7.3). In this system, both vegetation and microbial community are expected to remove the nitrogen where the first acts as a

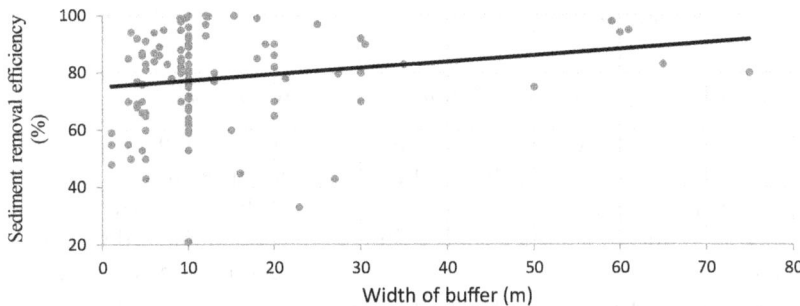

Figure 7.2 Summary of sediment removal efficiency of vegetated buffers (N = 114).

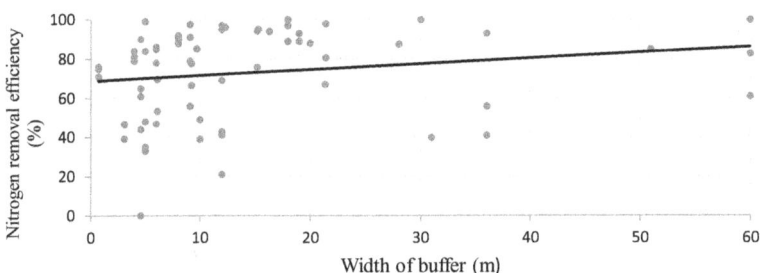

Figure 7.3 Nitrogen removal efficiency of vegetated buffers (N = 66).

temporary storage system whereas the latter facilitates the denitrification processes (Lowrance, 1992; Sabater et al., 2003).

7.3.5 *Efficacy of riparian buffers on nitrate removal*

Factors affecting removal of nitrates by riparian areas include the portion of flows crossing the riparian area as runoff, the rate of denitrification, and the time required for subsurface flows to cross the riparian area (Fennessy and Cronk, 1997). Ecohydrologically, denitrification is an important mechanism of nitrate removal (Brinson et al., 2002; Fennessy and Cronk 1997; Naiman and Decamps 1997) that in turn depends on the residence time of surface and subsurface water in response to width of the riparian area, slope gradient, surface roughness, hydraulic head, and soil hydrologic connectivity (i.e., permeability) (Spruill, 2000). In this regard, the average buffer length of 45 m is found to remove about 72% of nitrate [N = 54] (Figure 7.4).

7.3.6 Efficacy of riparian buffers on phosphorus removal

As phosphorus primarily enters buffers attached to sediments or as organic material (Wenger, 1999), the role of the buffer in reducing inputs to water bodies and wetlands conforms to the same mechanisms as that of reducing sediment inputs in general. Experimental research has demonstrated that even narrow grass buffers have the capacity to reduce phosphorus inputs. In this regard, the review result shows that an average buffer length of 14 is found to remove about 65% of phosphorus [N = 55] (Figure 7.5).

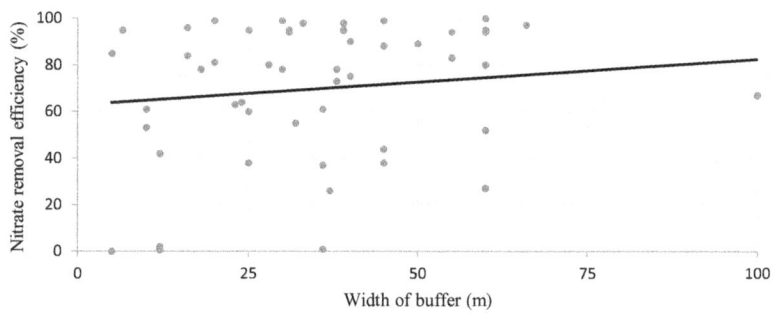

Figure 7.4 Nitrate removal efficiency of vegetated buffers (N = 54).

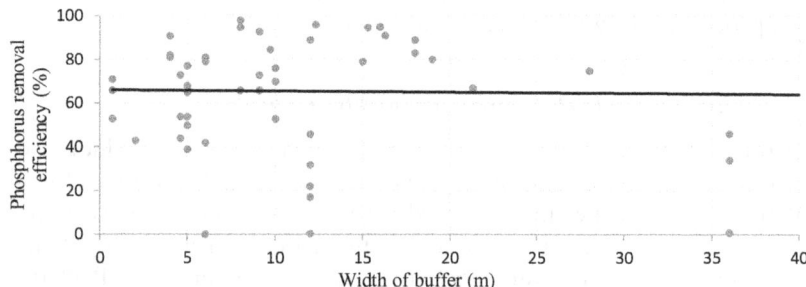

Figure 7.5 Summary of phosphorus removal efficiency of vegetated buffers (N = 55).

7.3.7 Ratio among trees- shrubs -grasses combinations for effective three-zone buffering

Because of the different modes of particulate and dissolved contaminant transport, multitier or combination buffers are often advocated (Fenemor and Samarasinghe, 2020). In theory, sediments and nutrients in surface runoff flowing from agricultural fields or timbered areas are intercepted first by the grass zone while nutrients entering deeper subsurface pathways are taken up by shrub and tree roots (NRC, 2002). Table 7.1 presents recommended ratio among trees, shrubs, and grasses.

7.3.8 Regression equations [buffer width vs. efficacy] from previous studies

Some literature attempt to estimate efficacy of vegetated buffer zone with buffer width by using regression equations as presented in Table 7.2.

7.4 Conclusion and recommendations

This meta-analysis provides the updated scientific information on the use of riparian buffer zone as a last line of defense for EcoLaR efforts. The information herein assists the planning as well as implementation of environmental management interventions. In conclusion, it is evident that the buffer width as well as the combination of vegetated zones differs according to the associated targets of the interventions. Consequently, it appears that prescription of fixed buffer width and setbacks may not be practical.

Table 7.1 Results of species diversity

Trees [Zone 1]	Shrubs [Zone 2]	Grasses/forbs [Zone 3]	References
Min 25 ft	Min 50 ft	Min 25 ft	
Min 25 ft	Min 55 ft	Min 20 ft	Schueler et al.
10m	30m	5m	(1992), Welsch
25 ft	50-100ft	25 ft	(1991)
15 ft	60ft	20 ft	

Table 7.2 Buffer width vs. efficacy

Sediment trapping efficiency (y) vs. buffer width (x)		References
1	$y = 0.0719 \ln x + 0.6783$ with $[R^2 = 0.27]$	Grass buffer strip	Yuan et al. (2009)
2	$y = 0.1179 \ln x + 0.5599$ with $[R^2 = 0.1]$	Forest buffer strip	
3	$y = 0.0714 \ln x + 0.6774$ with $[R^2 = 0.3]$	General	

TP removal efficiency (y) vs. buffer width (x)		
4	$y = 0.1155 \ln (x) - 0.0367$ $[R^2 = 0.7]$	Grass buffer strip	Cao et al. (2018)
5	$y = 3.6864 + 0.0436x$ $[R^2 = 0.45]$	Forest buffer strip	
6	$y = 0.1142 \ln (x) + 0.1432$ $[R^2 = 0.7]$	Grass/Forest	
7	$Y_p = 34.5 + 41 \log 10x_1 - 6.67\, dv_1 - 41.3\, dv_2$	dv_1 the first dummy variable for vegetation ($dv_1 = 1$ for forest and 0 for all others), dv_2 the second dummy variable for vegetation ($dv_2 = 1$ for bare ground and 0 for all others)	Bereitschaft (2007)

TN removal efficiency (y) vs. buffer width (x)		
8	$y = 0.1426 \ln (x) - 0.1938$ $[R^2 = 0.9]$	Grass buffer strip	Cao et al. (2018)
9	$y = 0.1108 \ln (x) - 0.2841$ $[R^2 = 0.66]$	Forest buffer strip	
10	$y = 0.1586 \ln (x) - 0.3401$ $[R^2 = 0.85]$	Grass/Forest	
11	$y = 10.5 * \ln (x) + 40.5$ $[R^2 = 0.14]$	General	Mayer et al. (2006)
12	$Y_n = 24.6 + 55.3 \log 10x_1 - 0.05\, x_2 - 14.4x_3$	x_1 the riparian width, x_2 the riparian percent slope, and x_3 the dummy variable for vegetation ($x_3 = 0$ if grass and $x_3 = 1$ if forest)	Bereitschaft (2007)

References

Abood, S., & Maclean, A. (2011). *Modeling riparian zones utilizing DEM, flood height data, digital soil data and national wetland inventory via GIS*. Milwaukee, WI: ASPRS Annual Conference.

Barfield, B.J., Blevins, R.L., Fogle, A.W., Madison, C.E., Inamdar, S., Carey, D.I., & Evangelou, V.P. (1998). Water quality impacts of natural filter strips in karst areas. *Transactions of the ASAE, 41*, 371–381.

Barling, R.D., & Moore, I.D. (1994). Role of buffer strips in management of waterway pollution: A review. *Environmental Management, 18* (4), 543–558.

Bereitschaft, B.J.F. (2007). *Modeling nutrient attenuation by riparian buffer zones along headwater streams* (Masters Thesis, Greensboro, NC: University of North Carolina).

Boyer, E.W., Alexander, R.B., Parton, W.J., Li, C., Butterbach-Bahl, K., Donner, S.D., Skaggs, R.W., & Grosso, S.J.D. (2006). Modeling denitrification in terrestrial and aquatic ecosystems at regional scales. *Ecological Applications, 16*, 2123–2142.

Brinson, M.M., MacDonnell, L.J., Austen, D.J., Beschta, R.L., Dillaha, T.A., Donahue, D.L., Gregory, S.V., Harvey, J.W., Molles, M.C., Rogers, E.I., & Stanford, J.A. (2002). *Riparian areas: Functions and strategies for management*. Washington, DC: Committee on Riparian Zone Functioning, National Research Council.

Burdon, F.J., Ramberg, E., Sargac, J., Forio, M.A.E., de Saeyer, N., Mutinova, P.T., Moe, T.F., Pavelescu, M.O., Dinu, V., Cazacu, C., Witing, F., Kupilas, B., Grandin, U., Volk, M., Risnoveanu, G., Goethals, P., Friberg, N., Johnson, R.K., & McKie, B.G. (2020). Assessing the benefits of forested riparian zones: A qualitative index of riparian integrity is positively associated with ecological status in European streams. *Water, 12* (4), 1178.

Cao, X., Song, C., Xiao, J., & Zhou, Y. (2018). The optimal width and mechanism of riparian buffers for storm water nutrient removal in the Chinese eutrophic Lake Chaohu Watershed. *Water, 10*, 1489.

Castelle, A.J., Johnson, A.W., & Conolly, C. (1994). Wetlands and stream buffer size requirements - A review. *Journal of Environmental Quality, 23*, 878–882.

Coats, R.N. (1999). *Riparian zone*. Dordrecht: Encyclopedia of earth science-environmental geology, Springer.

Daniels, R.B., & Gilliam, J.W. (1996). Sediment and chemical load reductions by grass and riparian filters. *Soil Science Society of America Journal, 60*, 246–251.

de Groot, R.S. (1992) *Functions of nature: Evaluation of nature in environmental planning, management and decision making*. Groningen: Wolters-Noordhoff.

Di Baldassarre, G., Viglione, A., Carr, G., Kuil, L., Salinas, J.L., & Blöschl, G. (2013). Socio-hydrology: Conceptualizing human-flood interactions. *Hydrology and Earth System Sciences, 17*, 3295–3303.

Dossey, M.G., Vidon, P., Gurwick, N.P., Allan, C.J., Duval, T.P., & Lowrance, R. (2010). The role of riparian vegetation in protecting and improving chemical water quality in streams. *Journal of the American Water Resources Association, 46*(2), 261–277.

Ekness, P., & Randhir, T.O. (2007). Effects of riparian areas, stream order, and land use disturbance on watershed-scale habitat potential: An ecohydrologic approach to policy. *Journal of the American Water Resources Association, 43*(6), 1468–1482.

Everett (2003). *Use of best available science in city of Everett buffer regulations: Non- shoreline streams.* Kirkland Washington: The City of Everett.

Fenemor, A., & Samarasinghe, O. (2020). *Riparian setback distances from water bodies for high-risk land uses and activities.* Landcare Research New Zealand Ltd and Tasman District Council, New Zealand.

Fennessy, M.S., & Cronk, J.K. (1997). The effectiveness and restoration potential of riparian ecotones for the management of nonpoint source pollution, particularly nitrate. *Critical Reviews in Environmental Science and Technology, 27*, 285–317.

Fischer, R.A., & Fischenich, J.C. (2000). Design recommendation for riparian corridors and vegetated buffer strips. Vicksburg, MS: EMRRP Technical Notes Collection. U.S. Army Engineer Research and Development Center.

Fortier, J., Gagnon, D., Truax, B., & Lambert, F. (2010). Nutrient accumulation and carbon sequestration in 6-year-old hybrid poplars in multiclonal agricultural riparian buffer strips. *Agriculture, Ecosystems & Environment, 137*, 276–287.

Gregory, S.V., Swanson, F.J., McKee, W.A., & Cummins, K.W. (1991). An ecosystem perspective of riparian zones: Focus on links between land and water. *Bioscience, 41*, 540–551.

Hansen, B.D., Reich, P., Cavagnaro, T.R., & Lake, P.S. (2015). Challenges in applying scientific evidence to width recommendations for riparian management in agricultural Australia. *Ecological management & restoration, 16*(1), 50–57.

Hawes, E., & Smith, M. (2005). *Riparian buffer zones: Functions and recommended widths.* Yale School of Forestry and Environmental Studies. For the Eightmile River Wild and Scenic Study Committee. New Haven, CT.

Hill, A.R. (1996). Nitrate removal in stream riparian zones. *Journal of Environmental Quality, 25*(4), 743–755.

Hruby, T. (2013). *Update on wetland buffers: The state of the science* (Final report. Washington State Department of Ecology Publication #13–06–11).

Jones & Stokes (2005). *Setback recommendations to conserve riparian areas and streams in western placer county* (pp. 03–133). Sacramento, CA: J & S.

Lam, Q.D., Schmalz, B., & Fohrer, N. (2011). The impact of agricultural best management practices on water quality in a North German lowland catchment. *Environmental Monitoring and Assessment, 183*, 351–379.

Leroux, S.J., & Loreau, M. (2008). Subsidy hypothesis and strength of trophic cascades across ecosystems. *Ecology Letters, 11*, 1147–1156.

Lind, L., Hasselquist, E.M., & Laudon, H. (2019). Towards ecologically functional riparian zones: A meta-analysis to develop guidelines for protecting ecosystem functions and biodiversity in agricultural landscapes. *Journal of Environmental Management, 249,* 109391.

Lowrance, R. (1992). Groundwater nitrate and denitrification in a coastal plain ripar-ian forest. *Journal of Environmental Quality, 21,* 401–405.

Luke, S.H., Slade, E.M., Gray, C.L., Annammala, K.V., Drewer, J., Williamson, J., Agama, A., Ationg, M., Mitchell, S., Vairappan, C.S., & Struebig, M.J. (2018). Riparian buffers in tropical agriculture: Scientific support, effectiveness and directions for policy. *Journal of Applied Ecology, 56*(1), 85–92.

Ma, M. (2016). Riparian buffer zone for wetlands. In C. Max Finlayson, M. Everard, K. Irvine, R.J. McInnes, B.A. Middleton, A.A. van Dam, N.C. Davidson (Eds.), *The wetland book*. Dordrecht: Springer.

Mander, U., Tournebize, J., Tonderski, K., Verhoeven, J.T.A., & Mitsch, W.J. (2017). Planning and establishment principles for constructed wetlands and riparian buffer zones in agricultural catchments. *Ecological Engineering, 103,* 296–300.

Mankin, K., Daniel, R., Ngandu, M., Barden, C.J., Hutchinson, S.L., & Geyer, W.A. (2007). Grass-shrub riparian buffer removal of sediment, phosphorus, and nitrogen from simulated runoff. *Journal of the American Water Resources Association, 43,* 1108–1116

Mayer, P.M., Reynolds, S.K., McCutchen, M.D., & Canfield, T.J. (2006). Riparian buffer width, vegetative cover, and nitrogen removal effectiveness: A review of current science and regulations (EPA/600/R-05/118). Cincinnati, OH: U.S. Environmental Protection Agency.

McKergow, L.A., Weaver, D.M., Prosser, I.P., Grayson, R.B., & Reed, A.E.G. (2003). Before and after riparian management: Sediment and nutrient exports from a small agricultural catchment, Western Australia. *Journal of Hydrology, 270,* 253–272.

Merrill, M. (2016). Riparian buffers: The lack of buffer protection policies and recommendations to expand protection. *Journal of Environmental Law and Litigation, 30,* 65–86.

Naiman, R., Bechtold, J.S., Drake, D., Latterell, J., O'Keefe, T., & Balian, E. (2005). Origins, patterns, and importance of heterogeneity in riparian systems. In G. Lovett, M. Turner, C. Jones, & K. Weathers (Eds.), *Ecosystem function in heterogeneous landscapes*. USA: Springer Science + Business Media, Inc.

Naiman, R.J., & Decamps, H. (1997). The ecology of interfaces: Riparian zones. *Annual Review of Ecology and Systematics, 28,* 621–658.

NRC (National Research Council) (2002). *Riparian areas: Functions and strategies for management*. Washington, DC: The National Academies Press.

Palone, R., & Todd, A. (1998). *Chesapeake bay riparian handbook: A guide for establishing and maintaining riparian forest buffers*. United States Department of Agriculture, Forest Service, Northeastern Area, State & Private Forestry. Natural Resources Conservation Service Cooperative State Research, Education, and Extension Service NA-TP-02–97, USA.

Phillips, J.D. (1989). Nonpoint source pollution control effectiveness of riparian forests along a Coastal Plain river. *Journal of Hydrology, 110*, 221–237.

Pinay, G., Susana, B., Benjamin, A.W., Anna, L., Eugenia, M., Francesc, S., & Stefan, K. (2018). Riparian corridors: A new conceptual framework for assessing nitrogen buffering across biomes. *Frontiers in Environmental Science, 52*, 3–18.

Quinn, J. (2003). *Riparian management classification for Canterbury streams* (NIWA Client Report: HAM-2003–064). Hamilton: National Institute of Water and Atmospheric Research Ltd.

Renouf, K., & Harding, J.S. (2015). Characterizing riparian buffer zones of an agriculturally modified landscape. *New Zealand Journal of Marine and Freshwater Research, 49*, 323–332.

Sabater, S., Butturini, A., Clement, J.C., Burt, T., Dowrick, D., Hefting, M., Maître, V., Pinay, G., Postolache, C., Rzepecki, M., & Sabater, F. (2003). Nitrogen removal by riparian buffers along a European climatic gradient: Patterns and factors of variation. *Ecosystems, 6*, 20–30.

Schueler, T.R., Kumble, P.A., & Heraty, M.A. (1992). *A current assessment of urban best management practices: Techniques for reducing nonpoint source pollution in the coastal zone*. Washington, DC: Metropolitan Washington Council of Governments.

Schultz, R.C., Isenhart, T.M., Colletti, J.P., Simpkins, W.W., Udawatta, R.P., & Schultz, P.L. (2009). Riparian and upland buffer practices. In H.E. Garret (Ed.), *North American agroforestry: An integrated science and practice* (2nd Ed.). WI: ASA, Madison.

Soykan, C.U., & Sabo, J.L. (2009). Spatiotemporal food web dynamics along a desert riparian-upland transition. *Ecography, 32*, 354–368.

Spruill, T.B. (2000). Statistical evaluation of effects of riparian buffers on nitrate and ground water quality. *Journal of Environmental Quality, 29*, 1523–1538.

Sun, G., Caldwell, P.V., & McNulty, S.G. (2015). Modelling the potential role of forest thinning in maintaining water supplies under a changing climate across the conterminous United States. *Hydrological Processes, 29*(24), 5016–5030.

Sweeney, B.W., & Newbold, J.D. (2014). Streamside forest buffer width needed to protect stream water quality, habitat and organisms: A literature review. *JAWRA Journal of the American Water Resources Association, 50*, 560–584.

Uusi-Kämppä, J., Turtola, E., Närvänen, A., Jauhiainen, L., & Uusitalo, R. (2012). Phosphorus mitigation during springtime runoff by amendments applied to grassed soil. *Journal of Environmental Quality, 41*. 420–426

Vidon, P., Allan, C.J., Burns, D., duval, T.P., Gurwick, N., Inamdar, S., Lowrance, R., Okay, J., Scott, D. & Sebestyen, S. (2010). Hot spots and hot moments in riparian zones: Potential for improved water quality management. *Journal of the American Water Resources Association, 46* (2), 278–298.

Welsch, D.J. (1991). *Riparian forest buffers: Function and design for protection and enhancement of water resources*. Radnor, PA: USDA Forest Service, Northeastern Area.

Wenger, S. (1999). *A review of the scientific literature on riparian buffer width, extent and vegetation.* GA: Office of Public Service and Outreach, Institute of Ecology, University of Georgia.

Wilcock, R.J., Betteridge, K., Shearman, D., Fowles, C., Scarsbrook, M.R., Thorrold, B.S., & Costall, D. (2009). Riparian management for restoration of a lowland stream in an intensive dairy farming catchment: A case study. *New Zealand Journal of Marine and Freshwater Research, 43,* 803–818.

Wilcox, B.P., Le Maitre, D., Jobbagy, E., Wang, L., & Breshears, D.D. (2017). Ecohydrology: Processes and Implications for Rangelands. In: Briske D. (Ed.), *Rangeland Systems.* Springer Series on Environmental Management. Cham: Springer.

Wilson, M., Boumans, R., Costanza, R., & Liu, S. (2005). Integrated assessment and valuation of ecosystem goods and services provided by coastal systems. In J.G. Wilson (Ed.), *The intertidal ecosystem: The value of Ireland's Shores.* Dublin: Royal Irish Academy.

Yuan, Y., Bingner, R.L., & Locke, M.A., (2009). A review of effectiveness of vegetative buffers on sediment trapping in agricultural areas. *Ecohydrology, 2* (3), 321–36.

Zalewski, M. (2014). Ecohydrology for engineering harmony in the changing world (pp. 1–18). In S. Eslamian (Ed.), *Handbook of engineering hydrology: Fundamentals and applications.* (1st ed.). Boca Raton, FL: CRC Press.

8 Framework for management strategies of natural wetlands

Mulugeta Dadi Belete

8.1 Introduction

'Wetland' is a collective term for marshes, swamps, bogs, and similar areas (Ramachandra, 2001) located at the transition between terrestrial and aquatic systems (Cowardin et al., 1979) whose formation, processes, and characteristics are largely dominated and controlled by water (Maltby, 1986). In other words, wetland is a halfway world between terrestrial and aquatic ecosystems and it exhibits some of the characteristics of both (Smith, 1980). Hydrologically, the water table of this ecosystem is at or near the surface of the land, or where the land is covered by shallow water (Ramsar, 2013). Globally, wetland as an ecosystem covers about 4–6% of the Earth's surface (Cools et al., 2013; Junk et al., 2013).

Although the value of wetlands for fish and wildlife protection has been known for centuries, some of the other benefits have been identified more recently. Wetlands are among the most productive ecosystems in the world, comparable to rain forests and coral reefs (Mitsch and Gosselink, 2015). On the contrary, they are fragile ecosystems that are susceptible to changes even with little change to the composition of their biotic and abiotic factors. They are also the least understood and most abused assets (Maltby, 1990); endangered ecosystems (Ramachandra, 2001); routinely overlooked (McInnes, 2013); underestimated (Turner et al., 2008); continually declining both in area and in quality due to increased anthropogenic pressure (Gebresllassie et al., 2014); and degraded beyond the socially optimal extent (Turpie et al., 2010). As evidenced by Cools et al. (2013) and Junk et al. (2013), 30–90% of the world's wetlands have already been destroyed or strongly modified resulting in more than US$ 20 trillion losses of ecosystem services annually (Gardner et al., 2015). The situation is likely to be more complex in developing countries. Their loss or impairment is usually

DOI: 10.4324/9781003309130-8

accompanied by irreversible loss in both the valuable environmental functions and amenities important to the society (Zentner, 1988).

Apparently, it is crucial to recognize that wetlands have critical ecohydrological role, so their proper management is of a paramount importance to ensure their integrity, resilience, productivity, and sustainability. Wetland management that generally refers to any positive activities on wetlands (Ramsar, 2010) is a relatively new field (Euliss et al., 2008). Planning process for their management needs to be holistic as well as systematic to give structure to and encourage a logical approach while considering a wide range of issues (Chatterjee et al., 2008) so as to improve efficacy of management efforts (Slocombe, 1993).

In terms of managing wetlands, there are diverse dimensions and multiple terms that are a bit complicated for easy take-up of the concepts by practitioners and managers of wetlands, especially in developing countries. For instance, Maltby (2009) perceived wetland conservation and management as two separate concepts while DEC (2012) treated the strategies as management and restoration. Ramachandra (2001) also conceived the concepts of management, conservation, and restoration differently. On the other hand, NRCS (2008) considered restoration, enhancement, and creation as the three management strategies of wetlands, while NRC (1992) reported the similarity of activities such as creation, re-allocation, and enhancement, to 'restoration' with some difference in the process of renewing native ecosystems to sites where they once existed. By recognizing such diverse naming and strategies of wetland management, this chapter synthesizes a general framework that will guide wetland managers. This framework is considered as one of the components of EcoLaR (ecohydrology-based landscape restoration) approach as wetlands are one of the integral parts of a landscape. The chapter also attempts to validate the designed framework to aid management planning by taking Cheleleka Wetland in the Ethiopian Rift Valley Lakes' Basin as a case study.

8.2 Ecohydrological functions of wetlands

Wetlands perform myriads of ecohydrological functions/ecosystem services (Figure 8.1) such as sediments removal (Phipps, 1986) to the rate of about 80–90% from run-off entering the system (Johnston, 1991); settling and filtration (Fennessy et al., 1994); removal of nitrogen through their biochemical processes such as nitrification and denitrification (Saunders and Kalff, 2001); and ammonia volatilization, ammonification, nitrification, nitrate ammonification, fixation, plant and microbial uptake, and ammonia absorption (Vymazal, 2006). They also remove dissolved phosphorus by absorption to soils and/or by precipitation with calcium to form calcium phosphate (Mitsch and Gosselink, 2000). Their ability to detain water also results in a natural die-off, and therefore removal from the water column, because many pathogenic bacteria cannot survive for long periods outside their host organism (Hemond and Benoit, 1988). In a similar way, wetlands reduce downstream erosion by slowing the velocity of water flowing downstream (Adamus et al., 1991) and limit flooding, moderate groundwater levels and base flow, assimilate nutrients, protect drinking water sources, and buffer coastal areas from storm surges.

Some of the mechanisms behind those ecosystem services by wetlands (Figure 8.1) include: phytodegradation (uptake contaminants into their root structure); rhizodegradation (biological degradation via plants secrete substances); and phytovolatilization (where contaminants entered the plants biomass and transpired through the plant leaves) (ITRC, 2003). Pollutants are also subjected to volatilization in wetlands that break down the compound and expel it into the air as a gaseous state (DeBusk, 1999).

Because of these vital ecohydrological functions, wetlands are recognized as living machines (McDonald, 1994); kidneys of the planet (Wallance, 1998); kidneys of the landscape (Mitsch and Gosselink, 1993); nature's kidney (Chatterjee et al., 2008); nature's water store on land (URT, 2014); biological supermarkets (Mitsch and Gosselink, 1993); wealth lands for various species of wild life (Lee, 1999); the most important ecological system on this planet and significant storehouses (sinks) of carbon and loaded with great genetic wealth (Lu, 2001); very good climate stabilizers; the most economically valuable and among the most biodiverse ecosystems in the world (Ramsar, 2018); worth of billions, perhaps trillions of dollars in ecosystem services (Schuyt and Brander, 2004); and sources of intangible spiritual and emotional well-being to peoples (IWMI, 2006).

8.3 Characterization and framing of the diverse wetland management strategies

As shown in Figure 8.2, this chapter characterizes and frames the diverse strategies of wetland management and coins the resulting framework as PREE (*Preservation; Restoration; Enhancement; and/or Establishment*). The PREE strategy of managing wetlands encompasses every possible option ranging from simple wetland preservation to more complex wetland creation.

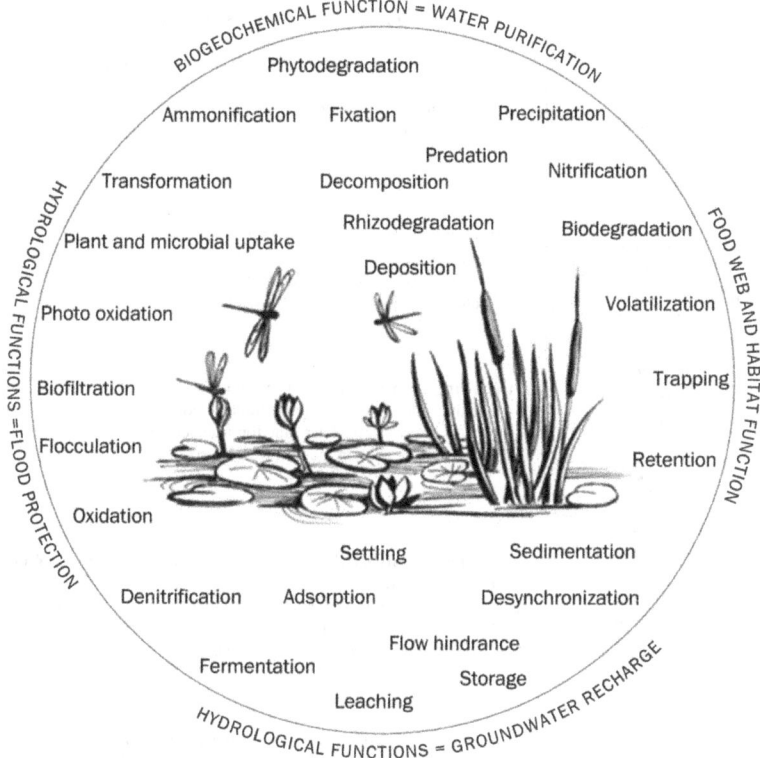

Figure 8.1 An example of a gully head with very vertical head in the middle of farmlands.
Photograph by the author.

It is considered that any given wetland management strategy can be categorized in one or more of the above four general categories. The following sub-sections describe each element of the PREE framework.

8.4 Preservation/protection/conservation/passive restoration (P)

This component of the strategic wetland management comprises avoidance /minimization /compensation /removal or prevention of the adverse anthropogenic pressures/threats facing the wetland (USEPA, 2021a). It can be achieved through avoidance of wetlands whenever

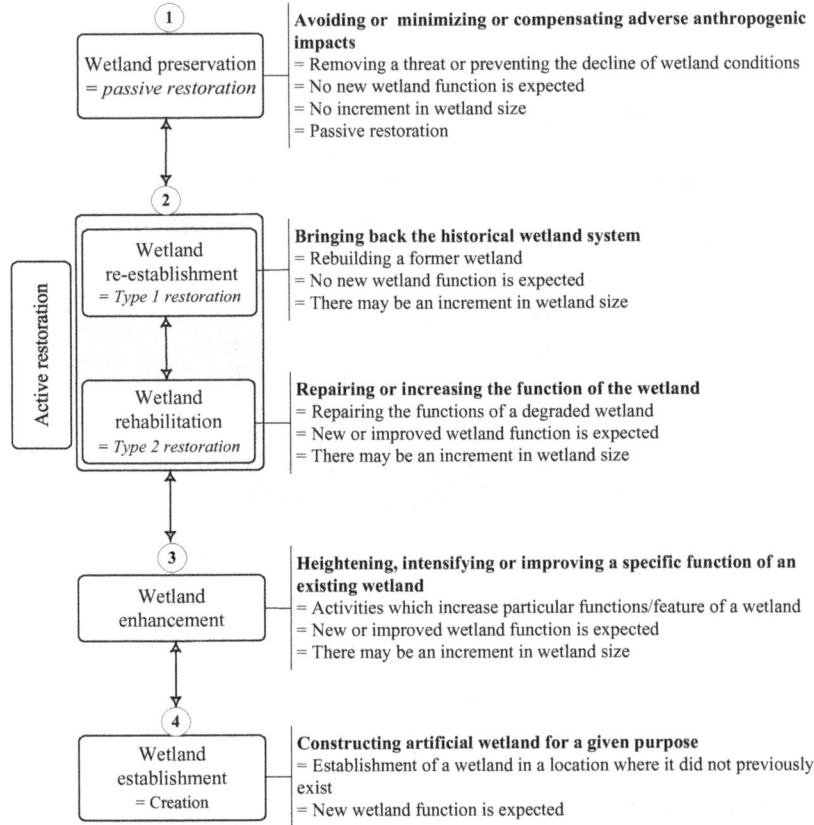

Figure 8.2 The PREE framework for strategic management of wetlands.

possible; or if you must disturb a wetland area, minimize disruption of the soil, vegetation and hydrology; or implementation of appropriate legal and physical mechanisms (i.e., conservation easements, title transfers). This strategy can also include protection of upland areas adjacent to wetlands as necessary to ensure protection or enhancement of the aquatic ecosystem. It can as well include passive restoration mechanisms such as creation of protected areas and protection of threatened species; preservation of certain types of natural habitat; integration of conservation of the natural environment within regional development and land-use legislation; identification, regulation, and management of processes which adversely affect biological diversity; and regulating activities that may adversely affect the conservation status of a given species (Shine and de Klemm, 1999). The preservation strategy does not result in a net gain of wetland acres and may only be used in certain circumstances, including when the resources to be preserved contribute significantly to the ecological sustainability of the landscape.

8.5 Restoration: re-establishment or rehabilitation (R)

In its generality, restoration of aquatic ecosystem implies the return of an ecosystem to a close approximation of its condition prior to disturbance with a general objective of emulating a natural, self-regulating system that is integrated ecologically with the landscape in which it occurs (NRC, 1992). This strategy includes 'Re-establishment' (the rebuilding of a former wetland) and 'Rehabilitation' (repairing the functions of a degraded wetland) (USEPA, 2007). The Restoration strategy targets a return of the natural or historic functions and characteristics of the wetland system (USEPA, 2000). By restoration, it can target restoration of ecological integrity (back to the condition of an ecosystem) – particularly the structure, composition, and natural processes of its biotic communities and physical environment. In other words, this strategy can be understood as a restoration of natural structure (to fix physical alterations) and/or restoration of natural functions (to re-establish the hydrological regime, natural disturbance cycles and nutrient fluxes). Oftentimes, restoration requires one or more of the following processes: reconstruction of antecedent physical conditions; chemical adjustment of the soil and water; and biological manipulation, including the reintroduction of absent native flora and fauna (USEPA, 2021b). Thus, restoration does not replace the need to protect aquatic resources in the first place, rather, it is a complementary activity that, when combined with protection and preservation, may result in a gain in wetland function or wetland size, or both.

8.6 Enhancement (E)

Enhancement (E) actions increase /modify /heighten /intensify /improve a specific function within the existing wetland system beyond the original natural conditions (Gwin, et al., 1999) that will cause more desirable conditions to prevail. This term includes activities commonly associated with the terms manipulation/directed alteration that involves 'making the good even better'. Again, it involves activities conducted within existing wetlands that heighten, intensify, or improve one or more wetland functions (Lewis, 1989). Enhancement is often undertaken for a specific purpose such as to improve water quality, flood water retention, or wildlife habitat. This strategy typically involves modification of site elevations or the proportion of open water (Gwin et al., 1999). Enhancement results in a gain in wetland function, but does not result in a net gain in wetland size.

8.7 Establishment (creation) (construction) of artificial wetlands (E)

The Establishment strategy denotes the development of a wetland or other aquatic resource where a wetland did not previously exist (Gwin, et al., 1999). In other words, creation occurs when a wetland is placed on the landscape by some human activity on a non-wetland site (Lewis, 1989) and is usually designed for a specific purpose that seldom has the full range of wetland functions and values provided by a natural wetland, even where they are of similar size (Shine and de Klemm, 1999) through manipulation of the physical, chemical, and/or biological characteristics of the site. Successful establishment results in a net gain in wetland size and function. Typically, this strategy includes excavation of upland soils to elevations that will support the growth of wetland species through the establishment of an appropriate hydrology (USEPA, 2021b).

8.8 Validation of the concept to appraise management plan of Cheleleka wetlands in the Ethiopian Rift Valley Basin

Once the PREE conceptual framework was devised, it was applied for the appraisal of management plan for Cheleleka Wetland, one of the wetlands in the Ethiopian Rift Valley Lakes' Basin. The site-specific management interventions were appraised in a participatory manner in the form of focus group discussion with the community elders and users of the wetland resources. Figure 8.3 presents the resulting management options and site-specific strategies that encompass three strategic management options for wetland Preservation; six for Restoration; two for Enhancement; and another two for Establishment strategy.

8.9 Conclusion and recommendations

This chapter has synthesized and framed the diverse components of wetland management strategies into a comprehensive framework known as PREE. Upon its synthesis and validation steps, the framework is recognized as a management tool for planning strategic

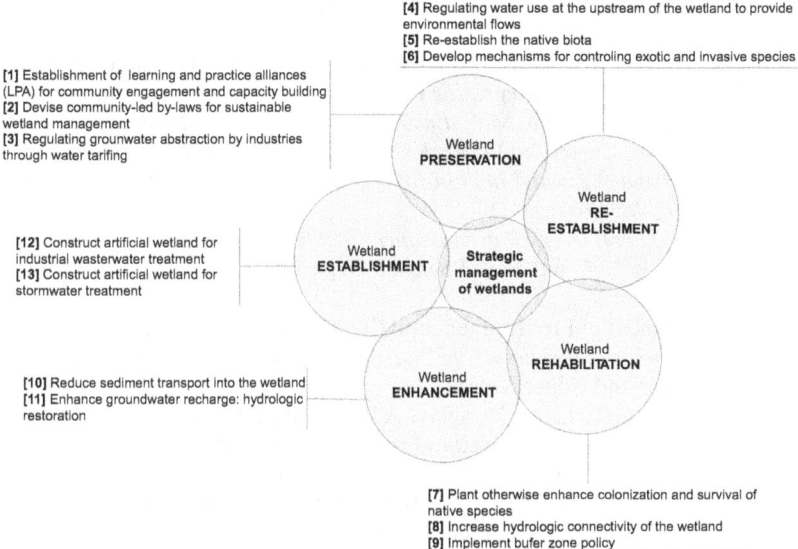

Figure 8.3 Validation result of the PREE concept for strategic management planning of Cheleleka Wetland (Ethiopia).

management of wetlands. As recognized from the framework, wetland strategies shall first attempt to 'do nothing' and the 'artificial/enginee ring/technological' solutions should be used as the last set of options. At the same time, it encourages synergism among the different interventions through optimum mix of management activities. It is also noted that the individual strategies are not mutually exclusive, and they are not procedural either. Rather, they are in the order of general preference and they can operate in synergy if exist simultaneously. It is safe to conclude that such approach potentially enables practitioners and wetland managers to plan sustainable management of their wetlands in a comprehensive manner. However, further development of the framework through subsequent researching will enhance the planning efforts in a more comprehensive manner.

References

Adamus, P.R., Stockwell, L.T., Clairain Jr., E.J., Morrow, M.E., Rozas, L.P., & Smith, R.D. (1991). *Wetland Evaluation Technique* (WET). Volume I: Literature Review and Evaluation. WRP-DE-2. Vicksburg MS: U.S. Army Corps of Engineers Waterways Experiment Station.

Chatterjee, A., Phillips, B., & Stroud, D.A. (2008). *Wetland management planning: A guide for site managers.* WWF, Wetlands International, IUCN and Ramsar Convention, India.

Cools, J., Diallo, M., Boelee, E., Liersch, S., Coertjens, D., Vandenberghe, V., & Kone, B. (2013). Integrating human health in to wetland management for the inner Niger Delta, Mali. *Environmental Science and Policy, 34*, 34–43.

Cowardin, L.M., Carter, V., Golet, F.C., & La Roe, E.T. (1979). *Classification of wetlands and deepwater habitats of the United States.* U.S. Fish and Wildlife Service, Washington, DC.

DeBusk, W.F. (1999). *Wastewater treatment wetlands: Contaminant removal processes* (SL155). Soil and Water Science Department, University of Florida, Florida.

DEC (Department of Environment and Conservation) (2012). *A guide to managing and restoring wetlands in Western Australia.* Perth: Department of Environment and Conservation.

Euliss, N.H., Smith, L.M., Douglas, A.W., & Bryant, A.B. (2008). Linking ecosystem processes with wetland management goals: Charting a course for a sustainable future. *Wetlands, 28*(3), 553–562.

Fennessy, M.S., Brueske, C.C., & Mitsch, J.W. (1994). Sediment deposition patterns in restored freshwater wetlands using sediment traps. *Ecological Engineering, 3*, 409–428.

Gardner, R.C., Barchiesi, S., Beltrame, C., Finlayson, C.M., Galewski, T., Harrison, I., Paganini, M., Perennou, C., Pritchard, D.E., Rosenqvist, A., & Walpole, M. (2015). *State of the World's wetlands and their services to people: A compilation of recent analyses* (Ramsar Briefing Note 7), Gland, Switzerland.

Gebresllassie, H., Gashaw, T., & Mehari, A. (2014). Wetland degradation in Ethiopia: Causes, consequences and remedies. *Journal of Environment and Earth Science, 4*(11), 40–49.

Gwin, S.E., Kentula, M.E., & Shaffer, P.W. (1999). Evaluating the effects of wetland regulation through hydrogeomorphic classification and landscape profiles. *Wetlands, 19*(3), 477–489.

Hemond, H.F., & Benoit, J. (1988). Cumulative impacts on water quality functions of wetlands. *Environmental Management, 12*(5), 639–653.

ITRC (Interstate Technology & Regulatory Council). (2003). *Technical & regulatory guidance for constructed treatment wetlands.* United States Environmental Protection Agency, Washington, DC.

IWMI (International water management institute) (2006). *Working wetlands: A new approach to balancing agricultural development with environmental protection* (IWMI Water Policy Briefing 21). Colombo: International Water Management Institute (IWMI).

Johnston, C.A. (1991). Sediment and nutrient retention by freshwater wetlands: Effects on surface water quality. *Critical Reviews in Environmental Control, 21*, 491–565.

Junk, W.J., Finlayson, A.S., Gopal, C.M., Mitchell, S., & Robarts, R.D. (2013). Current state of knowledge regarding the world's wetlands and their future under global climate change: A synthesis. *Aquatic Sciences, 75*(1), 151–167.

Lee, Y.J. (1999). Sustainable wetland management strategies under uncertainties. *The Environmentalist, 19*, 67–79.

Lewis, R.R. (1989). *Wetland restoration/creation/enhancement terminology: Suggestions for standardization* (The Status of the Science, Vol. II. EPA 600/3/89/038B. U.S. Environmental Protection Agency). Washington, DC: Wetland Creation and Restoration.

Lu, X. (2001). Wetland protection and water resources security in China. *China water resources, 11*, 26–27.

Maltby, E. (1986). *Waterlogged wealth: Why waste the world's wet places?* London: Earthscan.

Maltby, E. (1990). Wetland management goals: Wise use and conservation. *Landscape and Urban Planning, 20*, 9–18.

Maltby, E. (2009). *Functional assessment of wetlands*. Cambridge: Woodhead Publishing.

McDonald, L. (1994). Water pollution solution: Build a marsh. *Journal of American Forests, 100*(7/8), 26–30.

McInnes, R.J. (2013). Recognizing wetland ecosystem services within urban case studies. *Marine and Freshwater Research, 64*, 1–14.

Mitsch, W.J., & Gosselink, J.G. (1993). *Wetlands* (2nd ed.). New York: Van Nostrand Reinhold.

Mitsch, W.J., & Gosselink, J.G. (2000). The value of wetlands: Importance of scale and landscape setting. *Ecological Economics, 35*, 25–33.

Mitsch, W.J., & Gosselink, J.G. (2015). *Wetlands* (5th ed.). Hoboken, NJ: John Wiley & Sons.

NRC (National Research Counci) (1992). *Restoration of aquatic ecosystems: Science, technology and public policy*. Washington, DC: National Academy Press.

NRCS (2008). *Wetland restoration, enhancement, or creation*. Engineering Field Handbook (Part 650), Washington, DC.

Phipps, J.B. (1986). Sediment trapping in Northwest wetlands: The state of our understanding. In R. Strickland (Ed.), *Wetland functions, rehabilitation and creation in the Pacific Northwest: The state of our understanding*. Olympia, WA: Washington State Department of Ecology.

Ramachandra, T.V. (2001). Restoration and management strategies of wetlands in developing countries. *Electronic Green Journal, 1*(15), 1–18.

Ramsar (2010). *Managing wetlands: Frameworks for managing Wetlands of International Importance and other wetland sites*. Ramsar handbooks for the wise use of wetlands (4th ed., vol. 18.). Gland: Ramsar Convention Secretariat.

Ramsar (2013). *The Ramsar convention manual: A guide to the convention on wetlands* (6th ed.). Gland: Ramsar Convention Secretariat.

Ramsar (2018). *Wetlands- world's most valuable ecosystem-disappearing three times faster than forests, warns new report.* Retrieved from https://wwf. panda.org/wwf_news/?335575/Worlds-wetlands-disappearing-three-times-faster-than-forests.

Saunders, D.L., & Kalff, J. (2001). Nitrogen retention in wetland, lakes and rivers. *Hydrobiologie, 443*, 205–212.

Schuyt, K., & Brander, L. (2004). *Living waters conserving the source of life: The economic values of the world's wetlands.* Gland/Amsterdam: World Wildlife Fund.

Shine, C., & de Klemm, C. (1999). *Wetlands, water and the law. Using law to advance wetland conservation and wise use.* Gland, Cambridge and Bonn: IUCN,

Slocombe, D.S. (1993). Implementing ecosystem-based management. *Bioscience, 43*, 612–622.

Smith, R.I. (1980). *Ecology and field biology* (3rd ed.). New York: Harper and Row.

Turner, R.K., Georgiou, S., & Fisher, B. (2008). *Valuing ecosystem services: The case of multi functional wetlands.* London: Earthscan.

Turpie, J., Lannas, K., Scovronick N., & Louw, A. (2010). *Wetland valuation.* Volume I: Wetland ecosystem services and their valuation: A review of current understanding and practice (Report No. TT 440/09). South Africa: Water Research Commission. Cape Town: Republic of South Africa.

URT (United Republic of Tanzania) (2014). *Guidelines for sustainable management of wetlands.* Dodoma, Tanzania: Vice President Office.

USEPA (2000). *Principles for the ecological restoration of aquatic resources* (Office of water (4501-F). EPA841-F-00-003). USA.

USEPA (2007). *River corridor and wetland restoration.* Retrieved from http://www.epa.gov/owow/wetlands/restore/.

USEPA (2021a). *Types of mitigation under CWA section 404: Avoidance, minimization and compensatory mitigation.* Retrieved from https://www.epa.gov/cwa-404/types-mitigation-under-cwa-section-404-avoidance-minimization-and-compensatory-mitigation.

USEPA (2021b). Wetlands restoration definitions and distinctions. Retrieved from https://www.epa.gov/wetlands/wetlands-restoration-definitions-and-distinctions.

Vymazal, J. (2006). Removal of nutrients in various types of constructed wetlands. *Science of the Total Environment*, 380, 48–65.

Wallance, S (1998). Putting wetlands to work. *Journal of Civil Engineering, 68*(7), 57–59.

Zentner, J. (1988). Wetland restoration in urbanized areas: Examples from coastal California. In J.A. Kusler, S. Daly, G. Brooks (Eds.), *Urban wetlands.* Oakland, CA, Berne, NY: Proceedings of the National Wetland Symposium. Association of Wetland.

9 Synthesis and implication of the ecohydrology-based landscape restoration approach to the wider ecosystem management sector

The concluding chapter

Mulugeta Dadi Belete

9.1 'Synthesis' of the ecohydrology-based landscape restoration (EcoLaR) approach

In summary, the book is found to embed the concept of ecohydrology, which has developed into an extensive and recognized field in less than two decades (Zhou et al., 2016), in the landscape/ecosystem restoration concept. It starts with the philosophical foundation of the fairly new EcoLaR approach and describes the top ten limitations of the conventional practices in the landscape restoration sector. As a guide to the anticipated new path, the book sets out seven principles. Technically, the EcoLaR approach conceptualizes an innovative green-(semi) gray infrastructure as a kind of ecohydrological systemic solution. While presenting the contents of the green- (semi-) gray infrastructure in detail, it also presents the theoretical components and outcomes of field practices in the individual components of the infrastructure. The theory and practices of the EcoLaR approach demonstrated ways to address the key limitations of the conventional landscape restoration efforts. Some of the advantages of the new approach include: (1) maximized use of the role of dual regulation between 'biota and hydrology'; (2) highly reduced sizes of the physical structures; (3) the techniques avoid over-engineering of the environment with ecological integrity of the landscape; (4) the approach focuses on realization of water and nutrient cycle regulation; (5) once the hydrology is regulated, the EcoLaR approach maximizes the opportunities of the biota to deliver provisioning ecosystem services such as food production; (6) the approach is found to produce multiple ecosystem services as outcomes; (7) it is also found to provide immediate benefits to the community;

DOI: 10.4324/9781003309130-9

(8) the approach demonstrated the value of active restoration over the passive restoration attempts in significantly improving degree of stability of the landscape, degree of nutrient cycling, concentration of soil moisture phosphorus and potassium; (9) it is also demonstrated to be climate-smart through its flood regulation role as well as enhancement of soil moisture; and (10) the approach provides systemic solutions for the landscape restoration efforts.

Upon synthesizing the overall concept of the ecohydrology-based landscape restoration approach, it is concluded that restoration of ecosystems has never been more urgent (Ota et al., 2020) than at the moment and we are already late to step up our efforts (Young and Schwartz, 2019). This is evidenced by the declining trend of the ecological and regenerative potential of the planet Earth, expressed by the 'ecological footprint,' which recently is above 1.7 (EFP, 2021). This situation urges us not to continue in the state of 'business as usual' and escalates the interests in halting and reversing degradation and restoring landscapes (Stanturf et al., 2019). Zalewski (2021) pinpointed this situation as the point of regime shift from sociocentric/mechanistic paradigm into an ecosystemic one where the interventions are dictated by a profound understanding of how to regulate the fundamental ecological processes. These processes, first and foremost, concern cycling of water and nutrients and next, the whole range of water-biota interplay in different ecosystems which is a basic role of ecohydrology. This concept implies that restoration efforts are much more than simply planting trees and inevitably involve halting and reversing degradative pathways and creating transformative restoration systems that complement conservation and sustainable production systems (Chazdon and Brancalion, 2019). From this perspective, the EcoLaR approach is one of the recent research-based pathways that provides innovative solutions to increase the carrying capacity of degraded landscapes which is a much-expected outcome of the 21st-century problem-solving research.

When it comes to the transdisciplinary nature of the approach, most of the techniques in the EcoLaR approach are based on earlier concepts such as soil and water conservation, river engineering, landscape ecology, watershed management, ecological engineering, forestry, sustainable land management, stream restoration, ecology, and the like. Hence, the particularity of this approach lies in its capability to mimic important features of natural landscapes such as runoff-run-on system at hillsides as well as plunge pools and step-pool configurations in natural streams. In its simple form, the introduction of the two-phase mosaic of runoff-run-on system on hillslopes and the flow regulating barriers on agricultural lands can overcome a number of the limitations observed in the conventional practices of watershed management. They can do so through their capability to enhance water retention and quality, infiltration, and groundwater recharge. The effects of these techniques are especially significant in water-limited regions where the soil moisture deficit is generally high (Nahar et al., 2004). The techniques on farm lands can also provide alternative solution to farm bunds such as soil and fanya juu terraces which are among the most common techniques used in agriculture to collect surface run-off, increase water infiltration and prevent soil erosion (Waelti and Spuhler, 2021).

In this sense, the EcoLaR approach can be thought of being in line with the following saying:

> **To halt the decline of an ecosystem,**
> **it is necessary to think like an ecosystem.**
>> *Douglas P. Wheeler, EPA Journal, 1990*

Another essence of the EcoLaR approach lies in its capacity of tackling the challenges of landscape restoration through an integrated system of green- (semi-) gray infrastructure. The various components of the green- (semi-) gray infrastructure are expected to work together for effective regulation of the hydrology (*fulfilling the first principle of EH*) due to the fact that water is the common denominator and abiotic factor in shaping the biological structure and processes within ecosystems for all types of climatic, biogeochemical, and ecological processes (Zalewski, 2021). This principle conceptualizes the landscape as a 'superorganism' in a similar fashion as the Gaia concept of the planet as a 'superorganism' (Lovelock, 1995). Primarily, every landscape possesses a specific hierarchy of water cycle drivers related to the unique geomorphology, climate, plant cover, and so on (Zalewski, 2014). In any settings, the water cycling has to be considered as a primary regulator of ecological potentials such as bio-productivity and biodiversity (Wojtal-Frankiewicz et al., 2010). These abiotic factors need to become stable and predictable for the biotic interactions to start to manifest themselves (Zalewski and Naiman, 1985).

Following the hydrologic regulation in the landscape, which generally regulates biota, the second principle of ecohydrology views the 'superorganism' in a natural state as possessing resistance and resilience to stress. This principle recognizes the role of biocenoses in shaping the water and nutrient cycling as a fundamental tool to reverse the declining potential of the biogeosphere (Rodriguez-Iturbe, 2000; Tilman 1999; Vorosmarty and Sahagian 2000). It also embeds the analysis of ecological structure of the landscape in order to create a comprehensive plan for landscape restoration in terms of protection, rehabilitation, and management. Based on the first two steps, the third principle of ecohydrology (Ecological Engineering) leads us toward low-cost and high-efficiency nature-based solutions for landscape restoration (Zalewski, 2002) which is suggested to be as much an art as it is a science (Cooke et al., 2019).

9.2 EcoLaR approach in re-shaping the 'theory of change' of the contemporary landscape/ecosystem restoration initiatives

It is now well recognized that none of the 17 Sustainable Development Goals can be achieved unless a significant effort in ecosystem restoration is implemented (UN, 2021). In line with this necessity, a huge amount of finance and technical assistances have been pledged around the globe. Some of the many international commitments in the sector include: CBD Aichi Target 15; the UNFCCC REDD+; the Rio +20 land degradation neutrality; the Bonn Challenge; and most recently, the UN Decade on Ecosystem Restoration. However, the path toward achieving these ambitious targets remains unclear with some of these initiatives (Erbaugh and Oldekop, 2018; Stanturf et al., 2019). Through the EcoLaR approach, it is believed that these initiatives can fine-tune their intervention logics (theory of change) toward achieving their specific goals of natural resources management in general and of landscape restoration in particular. In other words, the likelihood success of these initiatives is firmly rooted in the underlying theory of change and actions.

Figure 9.1 indicates the likely pathways of achieving the generic goals of landscape restoration initiatives through the use of EcoLaR approach. As illustrated in Figure 9.1, the impact and the long-term outcomes are commonly shared among the contemporary initiatives of the restoration sector. However, the particularity of the EcoLaR approach appears at the level of short-term outcomes as well as the strategies. The short-term outcomes correspond to the success of the three principles of ecohydrology together with the realization of community stewardship. On the other hand, the strategies are combinations of mutually reinforcing and sometimes overlapping activities that tend to correspond with the EcoLaR guiding principles.

9.3 Conclusion

The general attempt of the EcoLaR approach to restore ecohydrologic relationships in landscapes is found to provide a fairly new

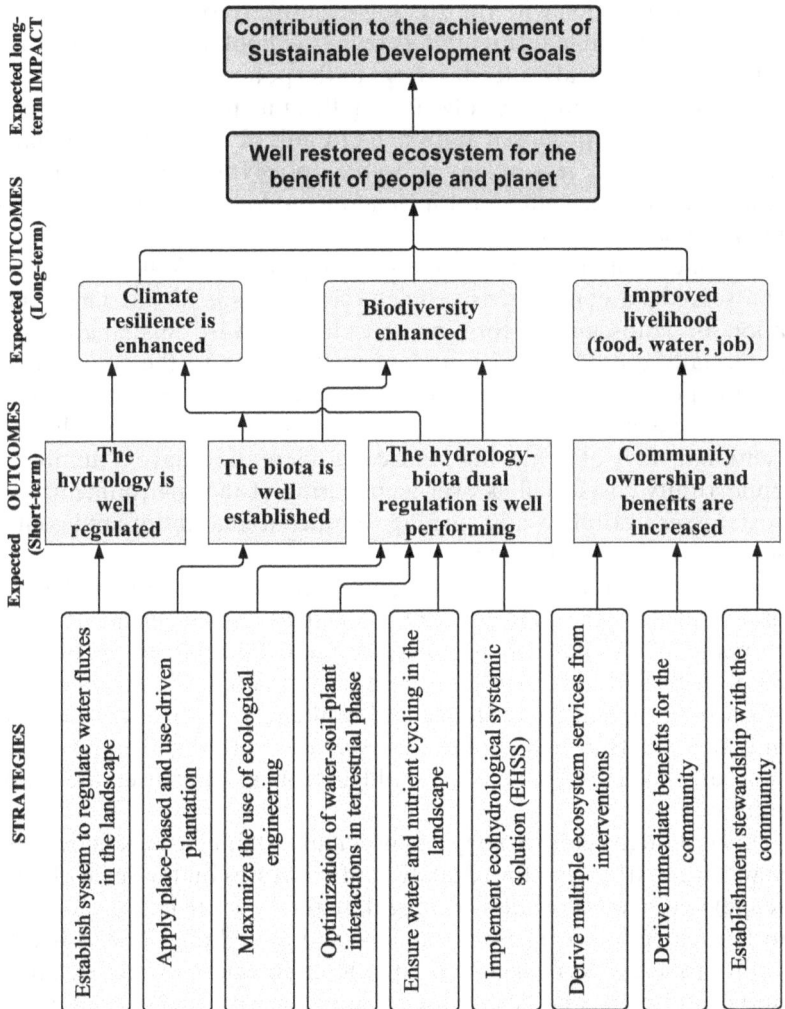

Figure 9.1 The proposed generic 'theory of change' for landscape restoration initiatives according to EcoLaR approach.

philosophy and methods of landscape level natural resources management in general and landscape restoration sector in particular. The approach is particularly important to restore degraded landscapes in water-limited conditions. However, it is not a stand-alone suitable approach for every degraded ecosystem. Nevertheless, it is a useful approach that can be linked to other synergetic efforts in the sector, and its combination with the existing scientific resources is expected to create a sustainable future. On its own, the book highlights the theoretical and practical understandings of the potential application of the ecohydrologic strategy to solve the challenging problems of landscape restoration. The proposed approach of EcoLaR issues positive signals for practitioners, researchers as well as policymakers in the natural resources management sector in general and for landscape restoration in particular.

In conclusion, it is recognized that the proposed EcoLaR approach can serve as a comprehensive scientific precursor for the new generation of landscape restoration paradigm toward sustainability. It potentially provides a cost-moderate alternative to the conventional landscape restoration initiatives while improving most of the limitations of the current efforts. It has also been recognized that the conventional ways of achieving landscape restoration have a number of opportunity costs such as over-engineering of the environment, high cost, lesser flexibility, and turning significant amount of lands out of production due to their physical structures.

Another lesson from the EcoLaR approach is that the underlying philosophy of the conventional watershed management strategies in many of the developing countries shall be revisited in order to align them with the pressing need of sustainability by increasing the 'absorbing capacity' (resistance and resilience) of ecosystems against human impacts, instead of mere conservation of the soil and water resources, which tend to view the problem from single causes and simple solutions.

Even though ecohydrology has new and old dimensions, it defines a new 'target' (the need for regulation of processes on the scale of landscape) and new 'know how' (to regulate processes from the molecular to landscape level). These are crucial in highlighting how to use the ability to adapt terrestrial and aquatic organisms to water dynamics in the landscape, and how to use the understanding of such feedbacks for the reduction of hydrological extremes (floods, droughts) and water quality improvement (Zalewski, 2010) that potentially shapes the future initiatives in the natural resources management sector.

9.4 Recommendations

This book is going to ignite professional dialogue on both theory and practices of adopting the principles of ecohydrology for effective landscape restoration. Consequently, more advanced and joint efforts of ecologists, hydrologists, and hydrological engineers will provide the scientific depth and width for sustainability in water resources management. Such advancement will essentially let the society go forward to opt for a transdisciplinary scientific approach that potentially formulates a concise and comprehensive vision and strategy for sustainability. In addition, both the theoretical as well as the practical dimensions of the EcoLaR approach are subjected to intensive improvements in response to experimental researches and stakeholders' feedbacks. Therefore, it is highly recommended that scientists in the area shall implement and experiment on the EcoLaR approach so as to strengthen its benefits and minimize the weakness toward realizing a sustainable environment around the globe.

References

Chazdon, R., & Brancalion, P. (2019). Restoring forests as a means to many ends. *Science, 365,* 24–25.

Cooke, S.J., Bennett, J.R., & Jones, H.P. (2019). We have a long way to go if we want to get the 'Decade on Ecosystem Restoration' right. *Conservation Science and Practice, 1*(8), e129.

EFP (Ecological Footprint) (2021). *Global footprint network.* Retrieved from https://www.footprintnetwork.org/our-work/ecological-footprint/.

Erbaugh, J.T., & Oldekop, J.A. (2018). Forest landscape restoration for livelihoods and well-being. *Current Opinion in Environmental Sustainability, 32,* 76–83.

Lovelock, J. (1995). *Gaia- a new look at life on earth.* Oxford: Oxford University Press.

Nahar, N., Govindaraju, R.S., Corradini, C., & Morbidelli, R. (2004). Role of run-on for describing field-scale infiltration and overland flow over spatially variable soils. *Journal of Hydrology, 286*(1–4), 36–51.

Ota, L., Chazdon, R.L., Herbohn, J., Gregorio, N., Mukul, S.A., & Wilson, S.J. (2020). Achieving quality forest and landscape restoration in the tropics. *Forests, 11,* 820.

Rodriguez-Iturbe, I. (2000). Ecohydrology: A hydrological perspective of climate-soil-vegetation dynamic. *Water Resources Research, 36*(1), 3–9.

Stanturf, J.A., Kleine, M., Mansourian, S., Parrotta, J., Madsen, P., Kant, P., Janice Burns, J., & Bolte, A. (2019). Implementing forest landscape restoration under the Bonn Challenge: A systematic approach. *Annals of Forest Science, 76*(50), 1–21.

Tilman, D. (1999). The ecological consequences of changes in biodiversity: A search for general principles. *Ecology, 80*(5), 1455–1474.

UN (2021). *Action plan for the decade on ecosystem restoration in Latin America and the Caribbean.* Bridgetown, Barbados: Forum of Ministers of Environment of Latin America and the Caribbean.

Vorosmarty, C.J., & Sahagian, D. (2000). Anthropocentric disturbance of the terrestrial water cycle. *Bioscience, 50*(9), 753–765.

Waelti, C., & Spuhler, D. (2021). *Bunds*. Retrieved from https://sswm.info/sswm-university-course/module-4-sustainable-water-supply/further-resources-water-sources-hardware/bunds.

Wojtal-Frankiewicz, A., Frankiewicz, P., Jurczack, T., Grennan J., & McCarthy, T.K. (2010). Comparison of fish and phantom midge influence on cladocerans diel migration in a dual basin lake. *Aquatic Ecology, 44,* 243–254.

Young, T.P., & Schwartz, M.W. (2019). The decade on ecosystem restoration is an impetus to get it right. *Conservation Science and Practice, 1*(e145), 1–3.

Zalewski, M. (2021). Ecosystem biotechnologies for the enhancement of eco-hydrological potential of the catchments - water, biodiversity, ecosystem services, resilience, culture and education. *IOP Conference Series: Earth and Environmental Science, 789,* 012031.

Zalewski, M. (2010). Ecohydrology for compensation of Global Change. *Brazilian Journal of Biology, 70*(3), 689–695.

Zalewski, M. (2014). Ecohydrology for engineering harmony in the changing world. In S. Eslamian (Ed.), *Handbook of engineering hydrology: Fundamentals and applications* (1st ed.). Boca Raton: CRC Press.

Zalewski, M. (2002). Ecohydrology-the use of ecological and hydrological processes for sustainable management of water resources. *Hydrological Sciences Journal, 47* (5), 823–832.

Zalewski, M., & Naiman, R.J. (1985). The regulation of riverine fish communities by a continuum of abiotic-biotic factors. In J.S. Alabaster (Ed.), *Habitat modifications and freshwater fisheries.* London: Butterworths, FAO.

Zhou, D., Zhang, H., & Liu, C. (2016). Wetland ecohydrology and its challenges. *Ecohydrology & Hydrobiology, 16*(1), 26–32.

Index

"*The Ethical Visions of Psychotherapy* is a much-needed contemporary analysis of the ways in which psychotherapy is inextricably tied to visions of optimal human functioning, of flourishing, of what makes life worth living, of the good life. Kevin Smith insightfully uncovers these implicit ethical assumptions even in those psychotherapeutic approaches that are purportedly nothing but technical applications of scientific findings. The book is thus an invitation to the often-neglected task of exploring how ethical and psychological strands interweave in psychotherapy."

Alan Tjeltveit, Professor of Psychology, Muhlenberg College, author of
Ethics and Values in Psychotherapy

"All too often our contemporary landscapes, be they international, national, social, ethnic, or professional, are torn asunder by relentless divisiveness and claims of rightness and superiority. Among the varied approaches to psychotherapeutic efforts, while the diversity of disciplines could well foster mutual learning and maturation, far too often the advocates of these models collapse into divisiveness and competitions that impoverish our opportunities to learn from one another. In these two volumes, Kevin Smith places ethics at the heart of these professional debates, examining and critiquing the values that divergent models of psychotherapy hold, both explicitly and implicitly, arguing that each represents a practice that promotes a particular vision of the good life. As a psychotherapist often drawn quite passionately into taking sides in these theory wars, I found in Smith's book a quiet, deeply resourced perspective that allowed me to take a more reflective stance with regard to both the differences and the commonalities of contemporary models of psychotherapy. These books will be of great value to practitioners, researchers, scholars, and teachers who value the reflective practice of the art, the science, and the philosophies of psychoanalysis and psychotherapy."

William F. Cornell, author of *Self-Examination in
Psychoanalysis and Psychotherapy*

"Kevin Smith's work is essential. Every single practitioner of psychotherapy should be familiar with Smith's message, and should be aware of the issues it raises for their work, every moment of every day. Psychotherapy, Smith tells us, is not a technical exercise in the amelioration of problems. The aims and conduct of psychotherapy are not adequately described or measured in the terms of evidence-based practice. Every aspect of psychotherapy, from the way problems are defined to the means by which they are addressed, is an expression, often inadvertent, of what we believe makes life good. Psychotherapy of every variety is a social practice, and like all such practices, it promotes an ethic. Whether they are used to thinking of their work as an ethical endeavor or not, all psychotherapists spend their entire professional lives influencing those with whom they work to live in certain ways and not in others. Psychotherapists are far too little aware of what is, after all, the very (ethical) ground under their feet.

Smith's two books should be assigned in every psychotherapy training program and should be required reading for those who have finished formal training, regardless of their theoretical orientation (yes, I do mean to include the entire spectrum, from psychoanalysis to cognitive-behavioral therapy) or the profession of its matriculants. Psychologists, psychiatrists, social workers, counselors, clergy, psychiatric nurses, marriage and family therapists—all really do need to think through the issues presented here."

Donnel B. Stern, Ph.D., William Alanson White Institute,
author of *The Infinity of the Unsaid*

The Ethical Visions of Psychotherapy

The standard view of psychotherapy as a treatment for mental disorders can obscure how therapy functions as a social practice that promotes conceptions of human well-being. Building on the philosophy of Charles Taylor, Smith examines the link between therapy and ethics, and the roots of therapeutic aims in modern Western ideas about living well.

This is one of two complementary volumes (the other being *Therapeutic Ethics in Context and in Dialogue*). This volume explores the links between therapeutic aims and conceptions of well-being. It examines several cognitive-behavioral and psychoanalytic therapies to illustrate how they can be distinguished by their divergent ethics. Smith argues that because research utilizing standard measures of efficacy shows little difference between the therapies, the assessment of their relative merits must include evaluation of their distinct ethical visions.

A key text for upper level undergraduates, postgraduate students, and professionals in the fields of psychotherapy, psychoanalysis, theoretical psychology, and philosophy of mind.

Kevin R. Smith is Adjunct Associate Professor of Psychology at Duquesne University, Pittsburgh, Pennsylvania, USA. He is a psychotherapist in private practice in Pittsburgh and supervises the psychotherapy training of psychiatry residents and doctoral students in clinical psychology. He has published papers on psychotherapy and phenomenological psychology.

Brian Schiff
Situating Qualitative Methods in Psychological Science

Brent D. Slife and Stephen Yanchar
Hermeneutic Moral Realism in Psychology: Theory and Practice

Jeff Sugarman and Jack Martin
A Humanities Approach to the Psychology of Personhood

Bethany Morris, Chase O'Gwin, Sebastienne Grant, Sakenya McDonald
Subjectivity in Psychology in the Era of Social Justice

Kevin R. Smith
The Ethical Visions of Psychotherapy

Kevin R. Smith
Therapeutic Ethics in Context and in Dialogue

https://www.routledge.com/psychology/series/TPP